Abraham Lincoln's DNA

and Other Adventures in Genetics

Abraham Lincoln's DNA

and Other Adventures in Genetics

Philip R. Reilly

COLD SPRING HARBOR LABORATORY PRESS
Cold Spring Harbor, New York • http://www.cshlpress.com

Abraham Lincoln's DNA
and Other Adventures in Genetics

Publisher	John Inglis
Project Coordinator	Mary Cozza
Production Editor	Patricia Barker
Interior Designer	Denise Weiss
Cover Designer	Ed Atkeson/Berg Design

Library of Congress Cataloging-in-Publication Data

Reilly, Philip, 1947–
    Abraham Lincoln's DNA and other adventures in genetics / by Philip R. Reilly.
        p. cm.
    ISBN 0-87969-580-3 (cloth : alk. paper)—ISBN 087969-649-4 (pbk: alk. paper)
    1. Human genetics—Popular works. 2. Medical genetics—Popular works.
    3. Genetic engineering—Popular works. I. Title.

QH431 .R38 2000
599.93′5—dc21   00-029467

20 19 18 17 16 15 14 13 12 11

All Cold Spring Harbor Laboratory Press publications may be ordered directly from Cold Spring
Harbor Laboratory Press, 500 Sunnyside Boulevard, Woodbury, New York 11797-2924. Phone: 1-800-
843-4388 in Continental U.S. and Canada. All other locations: (516) 422-4100. FAX: (516) 422-4097.
E-mail: cshpress@cshl.edu. For a complete catalog of all Cold Spring Harbor Laboratory Press publications, visit our World Wide Web Site http://www.cshlpress.com.

Preface

On a frigid January evening in 1972, as I was crossing the main quadrangle of Columbia University to the law school where I was a mediocre and discontented student, I had a flash of insight. It came at an important moment. Certain that I did not want to pursue a traditional legal career, I had been struggling for a year to find a bridge to another field. By temperament a generalist, I wanted to combine my legal training with studies in some other discipline and operate at an interface between the two. I hoped to do this in a novel and creative way. I had already flirted with psychology and anthropology, but long hours in the reading room at Low Library had dissuaded me from that direction.

The passage of 28 years has not blurred the moment I now recall. As I hurried through the darkness toward the lights of the law school, the words just seemed to pop into my brain. "Genetics. You should study genetics." It felt like a broadcast from some mysterious, far-off source. I have not since received such a simple, powerful directive from my subconscious.

As I pondered the idea during the next few weeks, and as I read about advances in human genetics (even then a favorite topic of science journalists), I became convinced. The early 1970s was the dawn of genetic engineering. Scientists were developing new tools that would permit them to dissect the DNA molecule, the stuff of which genes are made, in extraordinarily precise ways. Those tools would in time allow us to know ourselves at a more fundamental level than biologists or physicians had ever thought possible. Surely, I concluded, new insights about the structure and function of human genes, especially those that related to risk for disease, would raise profound questions for society and, thus, for law. So began a long journey which led, however circuitously, to this book.

Since those days at Columbia, I have spent countless hours thinking about the impact that advances in genetics are having and will have on society. I have usually framed these as legal or ethical questions. Does the state have the right to compel individuals to undergo mandatory genetic testing? Should a physician have a right, despite the objection of his patient, to warn close relatives about a serious genetic risk? Should the physician be liable for failing to warn? What rules should govern genetic research involving human subjects? Who may have access to archived human tissue for research purposes? Should the courts trying a criminal prosecution admit evidence that a defendant was born with a genetic predisposition to violence? Should all convicted felons have a DNA sample typed and stored in a databank, thus creating a genetic version of fingerprint files?

My interest in such questions took me on a decade-long journey. In 1973, after taking the bar exam, I became probably the first freshly minted lawyer ever to pursue graduate study in human genetics. After two years in the laboratory, I again changed course, spending a year as a fellow at Yale Law School. I then entered Yale Medical School, and the next seven years were dedicated to it and to a residency in internal medicine at Boston City Hospital. Despite the many wonderful experiences along the way, my goal never changed. I wanted to study human genetics and medicine for their own beauty, but I also hoped that the effort would give me a deeper sense of how advances in these fields might affect society. While I was pursuing this interest, the field of human genetics was transformed again and again by advances in molecular biology. By the mid-1980s it was clear that our ability to discover genetic facts about ourselves was going to surpass even my wildest speculations.

Today, as I redraft this preface, journalists around the world are writing articles about an extraordinary milestone. Working together, several scientific groups have completely sequenced human chromosome 22. Now and forever, we know that portion of the human blueprint encoded in the genes that reside there. Announcements about the completed sequencing of other chromosomes will appear ever more frequently. I imagine the completion of the last ten or so will not even stir much public interest, until of course, we have the entire 3,000,000,000 base pair sequence of the human species in hand. That will be cause for celebration!

The human genomic sequence has been hailed by some as the holy grail of biology. Decoding it will rank as one of the great intellectual achievements of our time. But this wonderful accomplishment has an Edenic feel. Are we competent to use genetic information in ways that confer far more good than harm? Do we even know where to begin? I am an unabashed champion of the value of genetic information, but I realize that the really extraordinary benefits of genetics will only become manifest if people learn something of the science. Today, only a tiny fraction of the population can honestly claim that it has done so. We must find new, effective ways to whet the world's appetite (especially among children) for learning about genetics. Human and medical geneticists have long paid lip service to this goal, but the evidence does not suggest that past efforts have converted many people to become lay students of genetics. This book is an experiment of sorts, a kind of dry run to see whether I can use stories to teach about genetics. I try to present genetic concepts and facts in ways that readers will barely notice, let alone find difficult or incomprehensible. If the next phase of my journey in genetics is to publish a book that will inform readers about some of the large public issues that flow from the successful decoding of the human genome and at the same time to teach them some basic science, so be it.

Philip R. Reilly

Introduction

We are poised on the brink of a fabulous milestone in human history. Sometime late in the year 2000 or in 2001, the world's newspapers will run a banner headline proclaiming that the final base pair (chemical letter) in the human DNA sequence has been identified and placed in its proper place on one of the 23 chromosomes. We have already sequenced the first billion bases, and the pace is accelerating. Today, no gene can elude us. Indeed, shortly after the completion of the consensus sequence (the 3,000,000,000 or so bits of DNA information that make up a haploid human genome, the DNA in an egg or sperm), it will be available on CD-ROM and easily downloaded from the Net. It will take many decades to decode this wonderful molecular book, but it will be worth the effort. For in reading the text, we will learn a great deal about the evolutionary history of our species and gain insights into how individuals interact with the environment.

Think of our wonderful complexity! Each of us has about 100,000 pairs of genes, itself a number large enough to impress. Somehow, these genes are self-organized to operate and maintain our trillions of cells. At any given moment, each cell in our bodies is performing under the guidance of a particular subset of these 100,000 genes, while the rest are quiescent. Much of the beautiful mystery of human embryology lies hidden in the program by which—over just a few weeks—genes turn on and off to create our hearts, our lungs, and our brains.

Because I have been trained in law, genetics, and medicine, over the last few years I have been asked hundreds of times to talk with groups of non-scientists about human genetics. The task is formidable: In an hour or two or three, teach some basic facts about genetics, provide an accurate description of our scientific powers, pose some issues that will drive home the immensely important relevance of genetics to our lives, and critique our early, bumbling efforts to deal with those issues. Since most of the people that I talk with (including physicians) have never studied genetics, and many lack confidence in their ability to learn science, I long ago realized that it is best to teach the subject in a painless way. I do this by telling stories.

There are really two books between these covers, one nested in the other. The obvious book is a collection of 24 stories about genetics arranged under six topics: history, justice, behavior, plants and animals, diseases, and ethical dilemmas. The historical figures include Abraham Lincoln, George III, and Nicholas II, the last Romanov tsar. Did Lincoln have Marfan syndrome? Should we try to find out? Does it matter? Did England lose its North American colonies because its king suffered from acute intermittent porphyria? DNA analysis has established to a certainty that a mass grave found in Yekaterinberg held the remains of Nicholas II and his family. What impact has that had on contemporary Russian society and on the Orthodox Church?

DNA evidence is having a profound impact on how we deal with crime. Today, the investigation of almost any crime assumes that there may be DNA evidence that will lead authorities to a suspect. To illustrate, I recount how a few cat hairs on a coat became the critical evidence in convicting the feline's master of murder. Impressed by the high recidivism rate among criminals, law enforcement officials have created a network of DNA databanks on convicted felons. In less than a decade, every state has set up, or has committed to set up, these banks, all of which will use a standard DNA testing technology developed by the Federal Bureau of Investigation. The public has paid little heed to this extraordinary development. How long will it be before the state routinely collects a DNA sample for identification purposes from all its citizens? Since we already conduct mandatory testing of infants for treatable genetic disease, it would be quite easy to save a drop of blood for such a databank.

Interest in using DNA evidence to solve crimes leads inevitably to the question of whether there are gene variants that predispose individuals to violent acts. This is an old fantasy. During the late 1960s, geneticists debated whether the presence of an extra Y chromosome predisposed men to violence. In the mid-1990s, researchers reported a family in which men with mutations in a gene that makes a brain chemical called monoamine oxidase were highly likely to commit violent crimes. How will our criminal justice system accommodate the discovery that certain people (albeit only a tiny fraction of all perpetrators) who commit crimes are biologically driven to do so? Will defendants someday be found to be not guilty by reason of genes? Will convicted felons undergo genetic testing as part of their evaluation for parole?

No field of science raises more troubling questions than behavioral genetics. With our new molecular tools we are already asking questions of immense societal significance: What role do genes play in predisposing a man or woman to schizophrenia or manic-depressive illness? Are there people in whom one could predict risk of such disorders? If a test were available, would you want your child tested to determine whether he carried an allele that predisposed him to schizophrenia? How would such knowledge affect how you and others perceived his potential or judged his mistakes?

If we succeed in finding genes that predispose to mental illness, can we hope to understand the role of genes in even more subtle topics such as the contours of personality? Geneticists are hard at work attempting to map genes that drive characteristic behaviors in different breeds of dogs. The results may present us with new and disconcerting insights into the impact that genes have on shaping human traits such as shyness and sociability. How might such information alter our understanding of human behavior? How might it affect theories of education? In 1994 scientists claimed that they had found a region of the X chromosome that contains a gene, a variant in which predisposes to homosexuality. Recently, similar research has refuted that finding. How strong is the evidence for a gay gene? What are the implications of the existence of such a gene variant? Should we even be investigating such questions? These are some of the issues that I explore in the chapters on behavior.

Part Four of the book looks at the impact that molecular genetics is having on our relationship with nature. Genetic engineering has given us a new dominion over the planet. Plant geneticists now transfer specific genes from one species to another almost at will. In the last few years, we have moved rapidly to end a chemical approach to controlling weeds and pests in favor of a genetic approach in which major crops are engineered to be resistant to a single powerful pesticide that eliminates the necessity for multiple sprayings. We are already growing millions of acres of genetically engineered soybeans. The yellow squash you will eat next week was probably grown from seed into which geneticists transferred a gene that confers resistance to the watermelon mosaic virus, which annually kills as much as a quarter of that crop. Many people have hailed genetically modified foods, but with growing zeal many more now vehemently oppose them. Do unknown dangers lurk in moving genes between species? Will feeding on genetically engineered corn pollen kill off the monarch butterfly? This is just one of many pressing societal questions that are exceedingly difficult to answer.

To feed 9 billion people (the projected world population in 2050), we will need to take a much greater percentage of our protein from the sea than we do today. Genetic engineering may well be the key to the "Blue Revolution." By tinkering with its gene for growth hormone, scientists have created salmon that grow to twice the usual adult weight during the first year of life. But there are unanswered questions. Have there been enough safety studies? Could we accidentally create superfish that, if they escaped their breeding pens, would forever change deep ocean ecology? I devote a chapter to discussing the advances and addressing the concerns.

Most of us readily agree that humans have failed to practice enlightened stewardship of the natural world, as is painfully obvious from the honor roll of species that we have extinguished. Genetic engineering raises exciting possibilities for repairing some of the damage by preserving endangered species, but it has raised a host of questions about the proper way to do so. Efforts to preserve the Florida panther, which I summarize in one chapter, force us to confront the ultimate preservation issue. Is it permissible to change a species to preserve it?

Nowhere is our growing power over nature more dramatically revealed than in xenotransplantation, the science of moving organs from one species to another. There is a major effort under way to genetically redesign pigs so that they can provide an inexhaustible reserve of needed hearts, livers, and kidneys. By moving human genes into pigs, we may be able to reshape the surface of their tissues so that our immune system will not recognize their organs as foreign when they are transplanted into our bodies. Ten years from now, several thousand people a year may avert kidney failure thanks to a genetically engineered organ harvested from a pig! What, if any, limits should be placed on xenotransplantation? Should we be able to do with cloned primates what we are currently trying to do with pigs? Does xenotransplantation carry a huge risk for humanity by permitting viruses that have for eons resided in pigs to take up residence in humans?

Twenty years ago, people thought that genetic diseases were rare, incurable disorders caused by mutations in genes that expressed themselves according to classic Mendelian principles of inheritance (dominant, recessive, and X-linked). Today, the term "genetic disease" is as likely to evoke thoughts about heart disease, mental illness, cancer, diabetes, or asthma, to name just a few of the many important disorders the onset of which is often influenced by a genetic predisposition. Advances in understanding the genetic component of human disease and how best to use that knowledge is a vast topic. To give the reader some sense of where we are headed, I have devoted chapters to one classic Mendelian disorder (cystic fibrosis) and two disorders for which in an important fraction of cases there is hard evidence of strong genetic liability—breast cancer and Alzheimer disease. The major focus here has been to explore the extremely difficult challenges of properly using genetic risk information.

About 1 in 25 white Americans carries a mutation in a cystic fibrosis gene. Scientists have designed extremely high quality, relatively low cost molecular tests to identify carriers. Should we provide universal premarital screening? By reviewing one's family history of breast and/or ovarian cancer one can estimate the likelihood that an individual will be positive if she undergoes DNA-based testing. Should the test be used more widely? How does one decide that question? How helpful is it to learn whether or not one is a carrier of a breast cancer gene or a gene variant that predisposes to Alzheimer disease? Should physicians or patients be in control of access to predictive testing?

Two of the questions that I have been asked most often in the last five years are (1) What is gene therapy? (2) When will it be available? People have been dreaming about somatic cell gene therapy—the correction of disease by delivering a normal gene to cells of affected individuals—for decades. Thus far, not a single cure can be claimed. Nevertheless, progress, especially in regard to developing effective ways to attach a payload of "healthy" DNA to viral rockets that will move on a biological trajectory to the nuclei of patients' cells, has been impressive. The tragic death in the autumn of 1999 of a young man after he had undergone gene therapy for a rare liver disease caused all involved in the field to reassess the status of our knowledge. I think it likely that we will develop effective gene therapies, particularly when the disease in question can be ameliorated by targeting a single, accessible tissue. For example, 10 years from now patients with cystic fibrosis might be treated effectively with "gene inhalers" (devices that spray a cloud of the normal version of the CF gene into the lungs), not unlike the way we treat asthma today.

The profound scientific and ethical questions in gene therapy arise when one contemplates germ-line genetic engineering—the alteration of germ cells to change the genetic constitution of an individual *and* his or her descendants. Thus far, scientists, religious leaders, and government policy wonks have all agreed that we should not undertake germ-line engineering. This is in part because they see it as a step toward genetic enhancement therapy, efforts to engineer embryos to be bigger, brighter, more musical, or whatever other dream one might have for one's kids. But interest in germ-line therapy will be impossible to contain if the technological hurdles are overcome. In 30 years or so, we will almost certainly have the capacity to genetically alter human embryos. What will this mean for society? Will it merely constitute the latest tool by which the upper tier of society maintains its economic lead over the lower quartiles? Or will it usher in deeper change?

In the last section of the book, I survey some key ethical dilemmas that have arisen and that will continue to complicate the implementation of advances in human genetics. At the moment, issues of genetic privacy are an overriding concern. State after state has enacted laws to limit the uses that health insurers may make of genetic information. Federal legislative interest is high. What is the crux of the issue? Who should have access to genetic information, and for what purposes may it be used? May a physician ever violate a patient's privacy to warn relatives about genetic risk? Under what conditions? Who decides? As genetic testing permeates medicine, will it change our ancient notion of confidentiality from one that is patient-centered to one that is family-centered?

After exploring the privacy debate, I take on two novel issues. During the 1990s there was growing concern for the moral and legal status of frozen human embryos. Throughout Europe and the United States there are tens of thousands of eight-cell human embryos suspended in tiny tubes immersed in liquid nitrogen. Although they were originally created to be implanted in an infertile couple, in many instances they are no longer wanted. What should be their fate if their "potential" parents die or divorce? Are frozen embryos people or property or something in between? Until quite recently, neither the parents nor the clinics had worked out a way to deal with such issues. How should their future be resolved when the couple from whose germ cells they were created divorce? How should the courts resolve such solomonic questions? I recount a fascinating, if painful, story about a divorce in which the only issue that divided the couple was control of seven frozen embryos.

No book about genetics could avoid discussing Dolly, the sheep created by cloning an epithelial cell from an adult sheep. How was this feat accomplished? How soon will we clone humans? What don't we know about Dolly? How old is Dolly? Although she was born in 1996, Dolly's progenitor DNA comes from an animal born in 1990. Rather than being young, she may be middle-aged! Is she fertile? Will she remain healthy? What threats, if any, does human cloning actually pose? Why is everyone so frightened by this prospect?

Throughout the last 50 years, genetics has labored under the shadow of the eugenics movement, a progressive idea that arose in late 19th-century England, took firm root in the United States, and became severely diseased in Nazi Germany. The old state-based negative eugenics programs—sterilization laws aimed at the mentally retarded, and immigration quotas targeted at those thought to be less genetically robust—have, thankfully, disappeared. But have they been replaced by a more subtle eugenics, one that is technologically enabled, physician-supported, and sought by couples as they plan their families? We now have the power to identify fetuses with birth defects in time to permit women to decide whether or not to abort them. As time passes, we will be able to assess fetuses with ever greater accuracy. What questions are the right ones to ask about human fetuses? What are the wrong ones? What does the advent of powerful screening tools that predict the future health or talents of individuals portend for how we view ourselves, our children, and fellow citizens with disabilities?

The other book, the one hidden inside the 24 stories, is a mini-genetics textbook. Every chapter contains important facts about genetics. For example, in the chapter on Abraham Lincoln I discuss a dominant disorder, Marfan syndrome, and introduce a powerful tool, the polymerase chain reaction. In the chapter on Toulouse-Lautrec I cover recessive disorders, consanguinity, and a rare genetic skeletal disorder called pycnodysostosis. In the material on mental illness I review the ups and downs of intense efforts to map a gene that predisposes to manic-depressive illness, research where claims of success have all too quickly given way to admissions of failure. To understand this story requires that one grasp the fundamental issues in gene mapping, a concept that is easy to master. I hope to teach some fundamental facts about genetics in a way that permits the reader to absorb them without effort.

The genetic revolution will be remembered as one of the great ascents of the human mind. We have started on a journey that will ultimately lead us to a world in which we will be able to influence our own evolution. This book, I hope, will give all who read it a deeper sense of where we are heading.

HISTORY

Using DNA to Understand the Past

President Lincoln with general George B. McClellan and group of officers, An-tietam, Maryland. (*Photo from Library of Congress American Memory Collection.*)

Abraham Lincoln
Did He Have Marfan Syndrome?

MARFAN SYNDROME

No one seems to know exactly how tall, but by all accounts Abraham Lincoln was an uncommonly tall man. During the Civil War a reporter described him as a "tall, lank, lean man considerably over 6 feet in height with stooping shoulders, long pendulous arms terminating in hands of extraordinary dimensions which, however, were far exceeded in proportion by his feet." Contemporary photographs confirm that he towered over most men. One famous photo shows him standing head and shoulders above the diminutive General McClellan, the caps of the other, taller officers just even with the beard on the President's chin. We know relatively little about Lincoln's health, but he was said to be a man of impressive strength and, except when he was chained by bouts of depression, of great energy.

The first person to suggest that Lincoln's height might be a sign that he had a genetic disorder known as Marfan syndrome was a Los Angeles physician. The idea arose by chance. In 1962, after diagnosing Marfan syndrome in a 7-year-old boy, the doctor traced the culprit gene through the family. He discovered that the little boy was an 8th-generation descendant of Mordecai Lincoln II, the great-great grandfather of the president. This by no means proves that Lincoln carried a copy of the gene that turned up in the child, but it is a tantalizing hint.

Marfan syndrome, which affects about 1 in 20,000 persons, is named for the French pediatrician who in 1896 first described a girl with severe skeletal abnormalities. Those who are born with this disorder may suffer from a wide variety of possible complications, all of which can ultimately be explained as due to defects in the connective tissue. The dislocation of the lens of the eye that sometimes occurs was reported in 1914, but it was

3

not until the 1940s that physicians realized that people with Marfan syndrome could die suddenly (and at a relatively young age) due to rupture of the aorta, the great vessel that carries blood from the heart. Because of the genetic defect in its tissue, decades of pounding by the surf of blood can eventually breach the vessel's wall, causing rapid death.

For a long time we could only speculate as to the cause of Marfan syndrome, but now we know. In July of 1991, three research teams simultaneously reported the discovery of the gene which, when defective, causes this disorder. Using a variety of cell staining and gene mapping techniques, the teams found that the responsible gene coded for a protein called fibrillin, one of the components of both the lens of the eye and the wall of the aorta. What clinched the proof was the discovery of two patients with no family history of the disorder yet who had both the classic clinical signs of Marfan syndrome *and* a mutation in the DNA in the fibrillin gene. These two people were "sporadic" cases, the result of a new mutation in either the mother's egg or father's sperm.

Like most genetic disorders, Marfan syndrome varies greatly in its severity. One important reason is that there are many spots in the gene where a mutation can occur, and some of these cause more damage to the protein that the gene is responsible for producing than do others. A second major reason is that it is caused by a defect in only 1 of about 100,000 pairs of genes which form the overall genome (genetic constitution) of an individual. Together, these other genes can either diminish or exacerbate the impact of the mutation in the fibrillin gene. It is possible that a man as apparently healthy as Abraham Lincoln could have been born with a mutation in the fibrillin gene and had a relatively mild form of Marfan syndrome.

THINKING ABOUT LINCOLN'S DNA

In the fall of 1990, as word was circulating through the research community that the gene responsible for Marfan syndrome had been pinpointed, Dr. Darwin Prockop, an authority on connective tissue disorders and the Director of the Institute of Molecular Medicine at Jefferson Medical College in Philadelphia, contacted the National Museum of Health and Medicine in Washington, D.C. to ask permission to have a tiny sample of Abraham Lincoln's preserved tissue for DNA analysis to determine whether he had in fact carried a mutation that can cause this disorder.

Like the Kennedy assassination, the circumstance of Lincoln's death is among the best-known stories in our nation's history. On the night of April 14, 1865, the President and his wife, accompanied by Henry Reed Rathbone, a trusted officer in the War Department, and his fiancee, were seated in a box at Ford's Theater watching the play, *Our American Cousin.* A single police officer, John F. Parker, was assigned to stand guard in the narrow hallway leading to the presidential box. At some point during the play, Parker left his post and took an empty seat in the theater. Taking advantage of this lapse, John Wilkes Booth, a second-rate southern actor who had carefully planned his attack, strode into the box and shot Lincoln once in the back of the head with a small derringer. To quote the autopsy report, the pistol ball traveled "obliquely forward toward the right eye, crossing the brain and lodging behind that eye. In the track of the wound were found fragments of bone which had been driven forward by the ball which was embedded in the anterior lobe of the left hemisphere of the brain."

Lincoln collapsed. Dr. Charles Leale, a young assistant surgeon who was the first physician to reach the President, felt no sign of life. But on finding the head wound, he promptly removed a blood clot with his finger, thus reducing the intracranial pressure, and Lincoln began to breathe. Four soldiers quickly carried the President out of the theater across the street to a rooming house owned by a man named Peterson. Dr. Robert Stone, Lincoln's personal physician, and Dr. Joseph Barnes, the Surgeon General, soon arrived and took charge. Through the night the cabinet members gathered and waited, helplessly. The efforts by the surgeons to remove the pistol ball failed, and all concluded that the wound was mortal. The President died the following morning at 7:22 A.M. Secretary of War Stanton is reported to have said, "Now he belongs to the ages."

It is not surprising that those present at the somber moment would realize that any artifact connected with the assassination would be of great historical curiosity. Dr. Leale, the young surgeon who first tried to help Lincoln, wrote that he had wandered through Washington in the early morning rain, and vowed to save his shirt cuffs that were stained with the President's blood. They have been kept by his descendants to this day. Many artifacts from that terrible night repose in The National Museum of Health and Medicine (in those days the National Army Museum) in Washington, D.C. They include the surgical instrument that Dr. Barnes used to probe for the pistol ball, the ball itself, two locks of hair (about 180 strands) from Lincoln's head, seven small fragments of his skull weighing about 10 grams, and the blood-stained cuffs of Dr. Edward Curtis, the pathologist who performed the autopsy. All are probably laden with Lincoln's DNA, the stuff of which genes are composed.

SHOULD WE TEST LINCOLN'S TISSUE?

DNA is a remarkably tough substance. Its long, double strands which sit inside the nucleus of cells can, if protected from the elements, last for millennia. With time the strands break and fray, but even short fragments can hold important information. Scientists now have incredibly powerful tools for isolating, amplifying, and studying the DNA from extremely tiny tissue samples, even a single cell. There is probably more than enough of Abraham Lincoln's DNA in the bone and hair to serve as a diagnostic sample, and it would be possible to extract it for study. This is what prompted Dr. Prockop to contact the museum. He speculated that he would need only a small piece of one bone fragment to obtain enough DNA to look for mutations in the fibrillin gene.

I became involved in the decision over whether or not to test Lincoln's DNA for evidence of Marfan syndrome because of Dr. Victor McKusick, the 1997 winner of the prestigious Lasker Award. For many years the physician-in-chief at the Johns Hopkins University School of Medicine and a founding father of modern human genetics, McKusick, a gentle soul with a hardy temperament, grew up on a farm in Maine. Although he was an identical twin (his brother was the Chief Justice of the Maine Supreme Court), Dr. McKusick doubts that this experience pushed him into genetics. He was an accomplished cardiologist before he turned to human genetics. Given that McKusick is the world's authority on Marfan syndrome and works in nearby Baltimore, it was inevitable that Dr. Marc Micozzi, the forensic pathologist who was at the time director of the National Museum of Health and Medicine, would seek his advice on how to handle Prockop's request. The two decided to convene a committee to advise the museum. Because of my background in both clinical genetics and law, Dr. McKusick invited me to serve.

On May 1, 1991, I joined a fascinating group that included Dr. Lawrence Mohr, one of the White House physicians; Cullum Davis, director of the Lincoln Legal Papers Project in Illinois; Cheryl Williams, president of the National Marfan Foundation; Lynne Poirier Wilson, a museum curator who is an expert on the management of special collections; and Colonel Victor Weedn, then chief of the Armed Forces DNA Identification Laboratory and responsible for the largest DNA bank in the world. We readily agreed that none of us had ever been asked to decide questions such as we now confronted.

Congressman John Porter (R-Illinois) had become concerned about the possibility that the museum would authorize the study of Lincoln's DNA, and he had formally requested that we answer four questions: (1) Is the proposal consistent with the best traditions of American scholarship and research? (2) Does the proposal violate Lincoln's privacy or his views on the disclosure of personal health and medical information? (3) Is it acceptable for a museum to allow specimens of great historic value to be destructively tested if a compelling public interest is served by doing so? (4) Is this proposal consistent with the prevailing standard of professional ethics in the disciplines of science and history?

The first, third, and fourth questions were relatively straightforward to address. Scholarly interest in the health of major historical figures and how illness may have influenced their behavior is a well-established area of research among historians. One need only think of the interest in the impact of his strokes on the presidency of Woodrow Wilson, the curiosity about whether John F. Kennedy was hampered by Addison's disease, and the fascination with how Franklin Roosevelt chose to deal with his disability. In 1980 pathologists published a detailed reanalysis of the histological slides and paraffin blocks containing part of a tumor that surgeons secretly removed from President Cleveland's palate in 1893, concluding that it was not an aggressive cancer. Lynne Wilson, our expert on the preservation of museum collections, reassured by the fact that DNA studies would consume only a tiny fraction of the holdings, concluded that the sacrifice of a tiny bit of bone would not harm the collection or compromise future scholarship.

The really challenging question was to try to determine what Abraham Lincoln would want us to do. Although Lincoln's is among the most studied lives in history, we had no firm historical information to guide us. The Lincoln experts on the panel knew of no action Lincoln had taken or letters he had written from which we could infer that he would either favor or oppose the proposed testing. The scholarly consensus was that Lincoln was not a particularly private person. Neither ethical nor legal analysis posed obvious roadblocks to testing. The law has long recognized that public officials and celebrities may not have the same expectation of privacy as do the rest of us. Furthermore, an individual's right to privacy dies with him, so there was no obvious prohibition to authorizing the test. The fact that there are no living direct descendants whose privacy might in some way be violated by a postmortem genetic analysis of their great ancestor's DNA somewhat simplified the issue.

Dr. McKusick and Cheryl Williams made the most compelling points in favor of testing. They argued that we live in a society in which many people with disabilities, including people with Marfan syndrome, suffer both overt and subtle discrimination. If it turned out that Lincoln, probably the most revered figure in our nation's history, had a genetic disorder, could not that fact be used in some way to strike a blow for human equality? At the least, could it not help people with Marfan syndrome, especially young people, bolster their resolve to deal with the trials, such as major surgery, that many of them must undergo? Although we could not be sure we were right, the panel members felt confident that we knew enough of his character to infer that if Lincoln were alive and learned that DNA testing of his blood might help another person, he would have readily consented to it.

Having resolved the ethical and curatorial issues, the panel reviewed the technical hurdles that would confront scientists if they tried to test Lincoln's tissue. In the spring of 1991, Dr. McKusick had consulted with two leading molecular biologists, Dr. Francis Collins, now the Director of the National Institute for Human Genome Research, and Dr. Uta Francke, a professor at Stanford. They were sure that the diagnostic challenges were substantial. The fibrillin gene is, as genes go, large, and at that time it was not yet well studied. Furthermore, it was possible that defects in one or several other genes that code for proteins involved in the structure of connective tissue could also cause Marfan syndrome. They advised that the wise course was to wait until molecular biologists had studied the DNA of several hundred living persons with the disorder so that they could decide whether a few common mutations caused most cases or whether each family burdened with the disorder had its own "private" mutation. It would be much more difficult to test Lincoln's DNA if there was no way to focus the molecular search. They had advised that the best course was to wait. We agreed, and decided to reconsider the matter a year later.

Later that afternoon, Dr. McKusick held a press conference in the office of Senator Mark Hatfield (R-Oregon), a respected Lincoln scholar. He described our panel's recommendation as a "qualified green light" for future testing. Newspapers across the country ran the story. Most people seemed to support the panel's decision, but not everyone agreed. One prominent bioethicist had already called the idea "voyeurism," and a political cartoonist argued that Lincoln would have said that biology was not important and that efforts to identify distinguishing biological facts ran counter to the search for political equality. A congressman, arguing that the panel had violated Lincoln's privacy, promptly drafted a bill to forbid testing, but it did not receive serious consideration.

The panel, now bolstered with several molecular biologists, met for the second and final time in Washington on April 14, 1992, the 127th anniversary of Lincoln's assassination. In the intervening year, a gene responsible for a Marfan-like syndrome called congenital contractural arachnodactyly had been located on chromosome 5. More important, studies of 28 families with the clinical diagnosis of Marfan syndrome had shown that in every one the cause was a defect in the fibrillin gene on chromosome 15. Studies of other families with rare disorders that mimic Marfan syndrome showed that, in them, the fibrillin gene was not the culprit.

The second meeting of the Lincoln panel was vastly different from the first, focusing exclusively on the technical feasibility of testing. To no one's surprise, as more families were studied, researchers were finding a steadily growing number of mutations. It would thus be impossible to do a conclusive test on Lincoln's DNA without studying the whole gene, a much more difficult task than looking for the presence or absence of a specific change known to be located at a specific molecular address. It was clearly not yet time to go forward with testing. Someone would first have to perfect techniques to extract a maximal amount of DNA from a small bone sample. Dr. Victor Weedn proposed that the technicians in his laboratory practice on a few of the many anonymous leg bones in the museum collection taken from Civil War soldiers who had undergone amputations. This would allow researchers to estimate how much DNA could be obtained from century-old bones. Because these bones had been boiled clean, their cells would have less DNA than would Lincoln's cells, but that would permit one to make a conservative estimate as to the amount of Lincoln material that would actually be needed.

An intriguing new issue emerged. What if a larger sample was needed than had been initially projected? There was a possible solution. Scattered across the country were families who claimed that they held garments stained with Lincoln's blood. Some of the claims, such as those of descendants of the soldiers who had carried Lincoln to the rooming house or the physicians who had attended him, were plausible. Why not use a bit of Lincoln's tissue held by the museum since his autopsy to create a DNA fingerprint, and then ask some of those families to submit a snippet of the bloody garments for DNA analysis? If the DNA profiles matched, it would confirm their claims, and it would identify the location of other samples for possible diagnostic studies. Dr. Micozzi agreed to look into the matter. The committee was unanimous that no testing be attempted until the fibrillin gene was thoroughly studied.

POSTSCRIPT

Nine years after the last meeting of the DNA Advisory Panel, we still do not know whether Abraham Lincoln had Marfan syndrome, and it may be a long time before we do know. The technological hurdles are still too high to guarantee success. If Lincoln had undergone a full autopsy, any recorded comment concerning the shape and condition of his aorta would have provided important evidence to support or reject the diagnostic debate, but the autopsy was limited to opening his head. Full autopsies did not become standard practice in medicine until the 1880s. Victor McKusick, whose guess is the best there is, gives the odds that Lincoln had Marfan syndrome at about 50-50, which, given the rarity of the disorder, is quite high. Someday we could know; DNA technology is advancing steadily, and our ability to learn a great deal from a small sample continues to improve.

The museum never went forward with the other aspect of testing—to learn whether preserved blood samples from around the country purported to be from Lincoln match a DNA profile created from tissue known to be derived from the President (because the samples have been in the museum since the assassination). In 1995, the brass at the Department of

Defense, the agency with ultimate jurisdiction over the museum holdings, squelched such plans. At the time, the department was being heavily criticized for the way it had organized its DNA bank to hold samples of all persons recruited into the military. Since the major public concern was that the privacy of the samples would be violated, officials feared that testing Lincoln's DNA would be seen as reinforcing that fear.

If testing Lincoln's DNA were to reveal that he had Marfan syndrome, it would be an interesting historical footnote. One could argue, for example, that Lincoln was therefore at risk for sudden death from a ruptured aorta, and that, even without the evil hand of John Wilkes Booth, he might not have lived through his second term in office or survived much beyond that to influence the course of reconciliation with the South. On the other hand, he was 56 when he died, and on the basis of his age and apparent vigor, there was no obvious suggestion of significant Marfan syndrome. He may have had a mutation, but only a mild form of the disorder.

A 1995 lawsuit provides a bizarre coda to the work of the Lincoln panel. Descendants of John Wilkes Booth petitioned the Baltimore Circuit Court to permit them to exhume the remains to obtain a DNA sample in order to determine whether the body was really his. This could be done by comparing DNA from the skeleton with DNA taken from the blood of living descendants. The move to open the grave was instigated by Nathaniel Orlowek, a Maryland high school teacher, and Arthur Chitty, a historian at the University of the South in Tennessee, who argue that the man that Union troops killed in Bowling Green, Virginia, 12 days after the assassination was not Booth. They concocted an argument that Booth escaped and lived out his life in Enid, Oklahoma, where he committed suicide 38 years later.

After hearing the arguments of the petitioners, the judge heard testimony by experts retained by the cemetery officials who opposed exhumation. James Hall, an historian and an author of a book on the assassination, described Chitty's theory as "utter nonsense." He said it grew out of a 1907 hoax by a man named Fennis Bates who wrote a book saying that a mummy he was selling was actually the body of Booth. James Starrs, a forensic expert, testified that it would be impossible to confirm that the remains were those of Booth by comparing DNA from a bone sample with DNA from the blood of a living relative unless one could locate a woman who was a direct descendant of Booth's mother through an uninterrupted line of female relatives. Starrs based his argument on the fact that to establish the familial connection, one would almost certainly have to analyze mitochondrial DNA, a special type of DNA, which, because it is present in eggs but not sperm, is transmitted through the ages from mother to daughter. The judge refused to authorize the exhumation, but the case could someday be reopened. Today, if a direct male descendant of Booth's father were alive, it would be possible to do this study by comparing DNA sequences located on the Y chromosome, material that is only transmitted from father to son.

Suppose that someday we do test Lincoln's DNA for Marfan syndrome. Will that set a precedent to ask another, perhaps more sensitive, question? The most serious health problem Lincoln bore in life was "melancholia," what we today call depression. There is ample evidence that he struggled with it off and on throughout his adult years. Lincoln bore many sorrows in life, especially the death of his children, including one son who died while Lincoln served in the White House. But his bouts of depression also arose at times when he was not buffeted by major personal loses. Could he have been genetically predisposed to depression? We know on the basis of family studies, including studies of identical twins reared apart and comparison of illness rates among adopted children and their biological and adoptive parents, that major depressive illness has a significant hereditary component. No single gene has yet been implicated, but the odds are high that in time we will identify several that may confer risk on those people who inherit mutations in them.

If someday we learn that Lincoln had a genetic form of serious depression, my estimation of him will, I am sure, rise even higher than it stands today. When I visit the Lincoln Memorial, as I often do, and glance at some of his great words carved in stone and think about the fearsome gale in which he was chosen to navigate the ship of state, I am awestruck. He saved the world's greatest democracy. To think that he might have accomplished that while struggling to fight off a thick gray fog of biochemical depression defies my imagination. Dr. Victor McKusick argues that a diagnosis of Marfan syndrome in Lincoln would help those who bear the disorder today. The same could surely be said for those with serious depressive illness.

Stamped out hundreds of billions of times on our pennies, looking gravely at us from the center of our five-dollar bills, gazing down at us in isolated majesty from his great memorial, Lincoln's face may be the most reproduced human image in American history. It is a face that we associate with justice and tenacity, a face that bore a nation's anger and its sorrow. His is a phenotype of greatness, regardless of the DNA sequence that shaped his body.

King Philip IV of Spain by Velázquez. Note the prominent lower lip and jaw.
(© The Frick Collection, New York.)

2

Kings and Queens
Genetic Diseases in Royal Families

MENDEL

The fundamental idea of heredity—that the germ cells transmit factors which shape basic characteristics in offspring and which are invariant through the generations—was discovered by a Moravian monk working alone in an Augustinian monastery in Brunn, which is today part of the Czech Republic.

Born in 1822 to a peasant family, Gregor Mendel must have been a top student from the start, for that alone could explain his climb from such humble origins. As was often then the case for bright men without other resources, the priesthood provided the path to a scholarly life. It was his high school physics teacher, Freidrich Franz, who recommended him to the Monastery of St. Thomas, saying, "In my own branch, he is almost the best." For a time after his ordination in 1848, he worked as a substitute teacher in village church schools. Curiously, although he failed an examination for a regular teaching license, his examiners recommended that he be sent to university for further study. After concentrating in mathematics and physics at the University of Vienna (1851–1853), Mendel returned to Brunn. He sat for the teaching license examination a second time but withdrew, possibly due to illness, which caused him again to receive a failing grade. He then settled into a routine life within the monastery, punctuated by regular work as a substitute teacher. In this he must have prospered, for in 1868 he was elected abbot. It was during the decade from 1854 to 1865 that Mendel, working in a small garden within the monastery, made his world-changing discovery.

The cornerstone of modern genetics, now a towering skyscraper of knowledge, is a single paper, first read by Mendel to his colleagues at the Natural Science Society of Brunn on the evenings of February 8 and

March 8, 1865, and published in its obscure journal in 1866. The journal had a subscription list of only 120, and the paper elicited no known effort to replicate its findings, a critical process for all scientific discoveries, for nearly 35 years. Mendel himself published only one other paper on plant breeding. In 1869, he reported that working with a plant called hawkweed (*Hieracium*) he had been unable to demonstrate the particulate nature of the hereditary material that he had shown so convincingly in his study of garden peas (*Pisum sativum*).

We have no idea why Mendel began his exhaustive study of heredity, but we do know why he chose garden peas. He wanted a hardy annual plant that was easy to grow, from which insects could not gather pollen to cause cross-fertilization, and that had a well-established number of strains with obvious physical differences. Easily available to him were strains that were tall or dwarf, strains with flowers that were white or red, strains with seeds that were round or wrinkled and green or yellow. For a decade Mendel bred hundreds of pea plants, crossing strains with one sharply contrasting characteristic against those with an opposite form. His genius was to follow them through the generations. He collected the seed from each hybrid plant, isolated it from the seed of all his other plants, and sowed it separately the following year. This allowed him to observe the characters associated with particular seed from particular matings over time.

Whenever Mendel crossed a tall plant with a dwarf plant, all of the offspring were tall. However, when he crossed the hybrid plants, he found that for every three plants that were tall there was roughly one plant that seemed to revert to the size of its dwarf grandparent. When he crossed a dwarf plant with a dwarf plant, the result was always dwarf offspring. From these experiments, repeated hundreds of times, Mendel deduced that in the seed of tall pea plants there was a factor that was transmitted intact through the generations and that was *dominant* to some corresponding factor in dwarf pea plants. From following crosses of tall plants he deduced that, despite their tallness, all of the hybrids must contain a factor which, if present in the seed of both parents, would yield dwarf plants, and that this happened in about one-fourth of the progeny. Because the tall plants masked the presence of the factor for dwarfness, Mendel called the dwarf factor *recessive*. Mendel repeated his breeding studies with six other easily characterized features, including flower color, seed shape, and seed color. In each case he was able to show that one of two characteristics was dominant over the other. All the work was summarized in his 1865 paper that announced the particulate nature of heredity and the concept of dominant and recessive traits.

In the 1920s, when geneticists reexamined his original data, they found that, given the large number of plants that he had studied, Mendel's numbers fit too well with the expected 3:1 ratio! This has led to accusations of fraud by Mendel or, in the alternative, suspicions that his assistant, knowing what Mendel thought he should be finding, rigged the count to satisfy his boss. But if Mendel's work is tainted, it can only mean that he somehow deduced the theory of particulate inheritance and then sought physical evidence to support it, which is even more impressive. For, unlike most scientific advances, there is no evidence that his discovery was guided by the work of others. There is simply no published literature on particulate inheritance remotely close to Mendel's work until his findings were rediscovered independently in 1900 by three botanists, an event that launched modern genetics.

After its rediscovery, the theory of particulate inheritance was investigated widely across many species, especially by animal breeders. The word "gene" was first used to describe a unitary, invariant, hereditary characteristic in 1906. By 1910 Mendelism, as it was often called, had refuted the notion of blending (that inherited characteristics represented a smooth mix of two sets of ingredients) that had dominated biological thinking in the second half of the 19th century.

Naturally, there was great interest in the application of Mendel's findings to humans. In the United States, Charles Davenport, Director of the Station for Experimental Evolution at Cold Spring Harbor, New York, was among the first and most forceful to apply the new theory of inheritance to humans. He and his contemporaries did this by compiling pedigrees. What Mendel had done by following the transmission of factors through generations, they did retrospectively. They looked for families with distinguishing features and tried to trace the features back through the generations. This was no easy task. Medical records were sparse, photography had only been available for about two generations, and they had little knowledge of what physical characteristics might reflect the underlying presence of single, dominantly acting genes.

DOMINANT GENES

In 1905, W. C. Farrabee, an American anatomist, reported the first example of dominant inheritance in humans—a large family in which about one-half of the members were born with brachydactyly (unusually short fingers). Of the descendants of a single affected woman born over four generations, 36 had the family hand. In 1917, Edward Drinkwater, an English pathologist, wrote up an astonishing confirmation of dominant inheritance in humans. He had examined a man with unusually short fingers due to a fusion of the first and second finger bones. The man, who reported that his father and grandfather had the same condition, was a lineal descendant of the first Earl of Shrewsbury, who was born in 1390. During repairs of the family burial vault, it was necessary to open the first earl's tomb, giving Drinkwater the opportunity to look at the skeleton. The long-dead earl also had markedly short finger bones, proving that this rare dominant gene had been present in the family for 14 generations!

Even better than earls for the early study of human genetics are kings and queens. Until about 1800, any unusual physical characteristics or odd medical problems were far more likely to be noted and recorded in these individuals and their families than in virtually anyone else in their era. Perhaps the earliest evidence of genetic disorders, as well as of the dangers inherent in incest, was discovered in the tombs of the pharaohs of Egypt. For centuries those dynasties practiced brother–sister and other close marriages. Because the pharaohs were considered to be gods, it was believed that no persons outside their blood line were a suitable match. Early in the 20th century, when European archaeologists opened royal tombs along the Nile, they found that many skeletons from the pharaonic lines showed evidence of congenital malformations, most likely, given the inbreeding, due to the effects of two recessive genes.

The most famous example of a dominant phenotype (the word, phenotype, refers to a discernible physical trait shaped by an underlying gene or genes), the result of one or many genes interacting with the environment, certainly the one most commonly used in genetic textbooks, has been provided by the Hapsburgs, the family that ruled the land that is now Austria from 1278 until the end of World War I. The ancient family name seems to have originated from the Habichtsburg or Hawk's castle, a small castle now in ruins in northern Switzerland, which was erected in 1020. With but one exception, Hapsburgs sat on the throne of the Holy Roman Empire from 1438 until it was abolished in 1806.

In the 15th century, a Bohemian princess married into the Hapsburg family, bringing with her a dominant gene that strongly influences the shape of the face. Perhaps the earliest evidence of the phenotype is a medallion showing the profile of Maximilian I (1459–1519). His distinct, narrow lower jaw and protruding lower lip are seen over and over through the generations. The Hapsburg face is unusually long, and the mouth tends to be partially open, an unattractive look, but one that court artists were apparently not directed to soften. Portraits of Emperor Charles V (1500–1558) and Archduke Albrecht (1817–1895) clearly show the distinctive face. The gene does not seem to have had any other effects; it certainly did not prevent its carriers from ruling the Austro-Hungarian Empire with iron fists for centuries.

George III

George III, the British king who lost the colonies, was often severely incapacitated by a dominant genetic disease, a rare disorder that is well understood today, but which was a complete mystery in the late 18th century. George III ascended the throne at the age of 22 when his father died suddenly on October 25, 1760. At the new king's side was John Stuart, the third Earl of Bute, a Scotsman who had risen to power during the 1740s as a favorite of Princess Augusta, the wife of the Prince of Wales, and the new king's mother. In the years immediately before his coronation, young George had been constantly under Bute's wing, and for years after, he viewed the political world through Bute's lens. In the early years of his reign, George's illness would give Bute tremendous power.

By all contemporary accounts, George III worked hard at learning to carry the mantle of kingship, but in the winter of 1765 he developed a strange malady that incapacitated him for the better part of three months. It was a second and much more severe attack of the same illness that had first come on after a cold in 1762. The symptoms were complex and baffling. At various times the king suffered from constipation, colic, chest pains, stomach cramps, skin lesions, an alarmingly fast pulse, profuse sweating, rapid and sometimes gibberish speech, swelling of the joints, loss of taste, gross irritability, hallucinations, delusions, and delirium. He also on numerous occasions passed urine which when left standing in a chamber pot was reported variously to turn crimson, purple, or the color of port wine. At times it was thought that he was near death. As his queen had not given birth to an heir, the illness of 1765 caused great political consternation, and led to the passage in May of a Regency Act, a law that laid out the succession to a king dying without issue.

The medical history, taken together with other contemporaneous records, clinches the diagnosis. George III suffered from acute intermittent porphyria, one of a group of six different disorders, each of which arises from a genetic defect in one of the six enzymes that are collectively responsible through a series of coordinated biochemical reactions for making a molecule called heme. Heme forms the core of hemoglobin, the molecule in red blood cells that carries oxygen from our lungs to our remotest cells. It is crucial to the respiratory cycle of cells. It is also the pigment that gives blood its color. Our bodies are constantly reprocessing old red blood cells, which are used up by the millions each minute. Cells in the liver normally use the breakdown products of hemoglobin to manufacture new heme.

People with the form of porphyria known as acute intermittent porphyria have a block at the beginning of this recycling pathway. Curiously, some people with the defective gene never become ill, whereas others become severely incapacitated. That is, the disease sometimes appears to skip generations. This strongly suggests an important role for environmental factors in triggering the illness. Scientists today think that symptoms arise when a chemical known as δ-aminolevulinic acid, which derives from the breakdown of hemoglobin, begins to accumulate. δ-Aminolevulinic acid is extremely toxic to the nervous system. An excess level was almost certainly the direct cause of the madness of King George III. One environmental factor that increases the risk of illness (probably by increasing levels of this chemical) in those who are born with the mutation is drinking alcohol. This is a particularly unfortunate fact, as brandy was one of the treatments ordered for the king when he suffered his attacks.

Although these different forms of porphyria each strike fewer than about 1 in 10,000 persons, the illnesses are well known. This may be due to the folklore that arose over the centuries in several isolated European villages where there were families burdened with a slightly different form of the disorder called congenital erythropoetic porphyria. Unlike the disease that incapacitated George III in his adult years, people with this disorder are affected from birth. They are extremely sensitive to sunlight, which blisters their skin. Over the years their skin and nails become brownish red due to the deposition of unmetabolized heme pigments, and they make blood-red urine. Because of the photosensitivity, affected persons, who also suffer psychiatric problems, naturally choose a nocturnal existence. One can easily see how, as the collective folk experience accumulated over the centuries, villagers could use their observations of these unfortunate individuals to create tales about vampires and werewolves.

Acute intermittent porphyria is known for the wide variability in the way it affects different people in the same family and even for its unpredictable expression in a single affected person. This was surely the case for the British royal family. The disorder is dominant, and it has been traced back from George III as far as Mary Queen of Scots, his grandmother six generations removed. Many biographers have described her behavior, especially her hopelessly foolish plot to murder her cousin, Elizabeth I, which led to her execution in 1587, as the work of a deranged mind. It is well established that Mary's son, James I, had acute intermittent porphyria, for there are records of him passing purple urine (the color is caused by the presence in the urine of unmetabolized iron-containing pigments). Yet, he does not seem to have been severely impaired. Nor were George III's great-grandfather (George I), grandfather (George II), or father (Frederick, Prince of Wales). George I lived to be 67 and was in relatively good health, and George II, who was also in good health, died suddenly of a heart attack at 77. Frederick, who died before his father, was taken suddenly at 44 by a lung infection (he was autopsied). Thus, a disease gene that had been traveling in the British royal family for at least eight generations had not caused really severe disease until it expressed itself in George III.

The diagnosis of George III was made, albeit somewhat late, in 1966 when two British physicians, Macalpine and Hunter, published the results of their detailed analysis of the many records kept concerning George's illness in the *British Medical Journal*. They report that during the worst bout of illness, which lasted for several months in 1788–1789, the king was committed against his will to a private asylum under the complete control of John Willis, a "mad doctor," a forerunner of today's psychiatrists. Willis, who, when the king was uncontrollable, sometimes constrained him with a straitjacket, and who placed too much faith in purgatives, generally used more benign techniques to calm and quiet his royal patient. Much time was spent in taking the king for country walks with his attendants, for example. We cannot know if these interventions helped, but they were far preferable to the treatments provided by the court physicians. The king, who had been periodically insane and sometimes near death since the attack started on June 11, 1788, began to convalesce rapidly in mid-February of 1789. During this eight-month period, one of the royal physicians, who closely followed his course, compiled some 40 volumes of handwritten notes, probably the largest medical record of the century.

George III was not again severely ill until 1801. During this attack, records show, he was clearly mentally impaired. He was unable to concentrate, often cried uncontrollably, engaged in many perseverative behaviors such as rolling handkerchiefs, and complained that he could not recognize himself in a mirror. Despite the best efforts of his physicians, who bled him and ordered that he drink brandy each night, he recovered. The king suffered yet another severe attack in 1804, but he again miraculously recovered. Unfortunately, his mind deteriorated and in 1810 he was adjudged to be permanently insane.

Despite being periodically ravished by his bizarre disease, George III lived to be 82, the longest reigning British monarch, save for Victoria. His last major attack of acute intermittent porphyria occurred in 1811. Interestingly, during his attacks of 1804, 1811, and when George III was dying in 1820, his oldest son, the Prince of Wales, with whom he had never gotten along, was also violently ill. In 1811, there was genuine concern that both men would die within days of each other. The Prince of Wales almost certainly also had acute intermittent porphyria, although in him it ran a less severe course.

Given the state of 18th-century medicine, it is hardly surprising that George III's physicians were of little help during any of the five major and many minor attacks that he suffered over the course of 50 years. The best of the royal physicians, realizing how little they knew, observed the course of the illness carefully and tried to protect the king by isolating him in a quiet rural setting when the illness exploded. The worst of them, especially during the major attacks, harmed him further with repeated bloodletting and worthless remedies such as drinking mare's milk.

There must have been countless occasions when his judgment was impaired by the effects of porphyria that George III was nevertheless either making decisions of great political importance or letting others exercise his authority. Of special interest is the impact of his extended illness in 1765 on the deteriorating relations with the North American colonies. We certainly cannot attribute the course of events that led to the Revolutionary War as being deeply affected by his illness, but George III may have for a time perceived his faraway subjects to be more intransigent than they really were, a perception that could not have helped hopes for reconciliation.

X-LINKED GENES

Mendel definitively described two of the three classic modes by which genes that travel through generations affect us—dominant and recessive (see Chapter 3). It would be another 50 years before the scientific basis for the third mode, sex-linked or X-linked inheritance, was clearly grasped. In humans the genes are arranged on 23 pairs of chromosomes, 22 pairs of autosomes and a pair of sex chromosomes. Women have two X chromosomes and men have one X and one Y chromosome. The Y chromosome transmits the gene that determines maleness, but compared to the much larger X chromosome, it has relatively few genes. The consequences of this for the two sexes are immense. If a girl inherits an X chromosome with a potentially harmful recessive gene, it is almost certain that the comparable gene on her other X chromosome will protect her from it. If a boy inherits an X chromosome with a harmful gene, there is no corresponding gene on his Y chromosome to counter its effects, and he will become ill.

Nearly two millennia separate the earliest record of humanity's practical understanding of X-linked disease genes from the scientific understanding of how they operate. By the 2nd century A.D., the Talmud had rules regarding the ritual circumcision of boys in families in which death had occurred as a result of excessive bleeding following the operation. The rules forbade the circumcision of later-born sons of a woman who had lost two boys from uncontrollable bleeding. In addition, it admonished that the sons of her sisters should not be circumcised. Yet, half-brothers of the dead sons, if sired by the same father and a different mother, were to be circumcised. Clearly, those who wrote these rules realized that there was a hereditary risk of bleeding that was carried by and passed through the mother. Further elaboration did not come until about 1820, when a German physician recognized that daughters of men with bleeding diseases, although themselves unaffected, could pass the disease on to their sons.

Only about 1910, as geneticists studied the inheritance of certain traits in animal breeding and as others began to map the location of genes on the chromosomes of fruit flies, did it become clear that the Talmudic observations could be explained by positing that a gene that affects how blood clots must be located on the X chromosome.

During the mid-19th century, as scientists were beginning to grasp the idea of sex-linked inheritance, another long-lived British monarch, Queen Victoria, was born with what is almost certainly the most famous mutation in history. Today the remnants of Europe's royal families include many persons with hemophilia. By reconstructing the pedigrees, it is readily apparent that all affected persons are descendants of Victoria, and that none of her ancestors were affected. Thus, either the egg from her mother or the sperm cell from her father must have carried within it a new mutation in the gene that codes for a protein essential to blood clotting that is known as factor VIII (the name dates from a time when scientists were working out the complex pathway by which blood clots, but before they had identified the enzymes that did the work). Victoria was a silent carrier who could not know the risk she posed to her children and grandchildren.

Hemophilia is the most common of the serious clotting disorders, affecting about 1 in 10,000 persons. As is true with most genetic disorders, the severity of the disease depends on which of the hundreds of different possible mutations are in the gene of the particular patient. The reason that there can be hundreds of different mutations is that the DNA sequence that codes for a protein typically is composed of thousands of chemical units called bases. A change in even one base can give rise to a defective protein. Depending on how much damage is done by the mutation, some people are at risk for serious bleeding only after major injury, some bleed severely in response to minor injuries, and some—the most severely affected—bleed spontaneously, especially into their joints, which in the past caused serious skeletal problems. Until the development of factor VIII therapy, the periodic transfusions of donor plasma enriched for the needed protein, children with severe mutations usually died.

The great progress with factor VIII therapy had a tragic setback in the 1980s when some children, especially in France where the blood supply was not sufficiently monitored, became infected with HIV and died of AIDS. Today, thanks to the tools of genetic engineering, we use bacterial factories to make vast amounts of pure factor VIII. The treatments pose little risk of an allergic reaction and no risk of HIV infection. Today, even children with severe hemophilia can be greatly helped by such therapy.

Unquestionably, the most famous recipient of Victoria's new and, unfortunately, severe mutation was her great grandson, Alexis, heir to the throne of Russia. Victoria passed the chromosome with this mutation to her daughter, Alice, who silently passed it on to her daughter, who grew up to marry Czar Nicholas II of Russia. When the empress discovered that Alexis had severe hemophilia and realized that contemporary medicine had little to offer him, she turned elsewhere for help. She came under the influence of Rasputin, a monk with allegedly mystical healing powers, and soon placed Alexis under his care. The boy seemed to improve, perhaps because Rasputin successfully used hypnosis to help him avoid even minor trauma. Impressed with Rasputin's work, the empress manipulated the tsar to heed the monk's advice on an ever-widening array of matters. For a short time, he was among the most influential persons in Russian politics. The decadence and corruption that Rasputin encouraged during his brief flirtation with power almost certainly helped to weaken the tsar's hold on his nation as it spiraled into revolution. It is intriguing to speculate how the course of modern European history might have differed had Victoria not been born with a mutation in her factor VIII gene.

The mutations that cause acute intermittent porphyria or hemophilia or Marfan syndrome (see Chapter 1) provide some of the most dramatic examples of the influence of a single mutation on the body. The knowledge of the impact of a few disease genes in the history of a few famous families helped formulate concepts that led to the much deeper understanding of gene action that we have today. Classic Mendelian disorders are the great peaks in the landscape of genetics, the landmarks that the early explorers could set their sights and their hopes on as they marched along. But the interaction of gene pairs, the biochemical dance of dominance and recessiveness, is by no means limited to genetic disorders. It is played out trillions of times, usually in extremely subtle ways, in our cells from the moment of conception until death. Today, as we shall see, molecular biologists can ask far more subtle questions about that minuet.

Henri de Toulouse-Lautrec. (© Georges Beaute.)

Toulouse-Lautrec
An Artist despite His Genes

In the summer of 1997, I had the good fortune to take my family on holiday to Paris. Among the small souvenirs we brought home was a set of coasters for drinks, each depicting a famous Parisian landmark. They were the Eiffel Tower, Notre Dame, Sacre Coeur, the Arc de Triomphe, Saint Chapelle, and the Moulin Rouge. That the last, a tawdry dance hall set up in the 1880s on the outskirts of Paris, should hold rank with the other five is in one sense a measure of the impact of a talented, unconventional, and tortured painter whose meteoric career lasted less than 20 years.

Fate spun out an unusually complex web for Henri Marie-Raymond Toulouse-Lautrec Montfa, who was destined to scandalize and then win over the art world with bold paintings and posters of the seamy side of life in Montmartre. Born in 1864 to great wealth, he was the first and only surviving son of an avid horseman and huntsman who lived a life unconstrained by social rules and a pious, retiring mother who could not have been more opposite. The night of his wife's delivery, her husband, Count Alphonse, is reliably reported to have prayed feverishly for a son, mainly so that he could train a future hunting companion. Famous for his wild life, Alphonse was a descendant of the ancient Counts of Toulouse and Viscounts of Lautrec whose ancestors had been the cousins of kings. Adele Tapie de Celeyrais, the future painter's mother, was from a less illustrious family, but her grandfather owned a magnificent estate with 3500 acres of vineyards. She was swept off her feet by her future husband's dashing military ways, and her parents, putting social class and economics ahead of character, encouraged a marriage that would soon turn sour. Crucial to our tale is that Henri's parents were first cousins.

Consanguinity and Recessive Genes

Long before geneticists began to do research, folk wisdom recognized that there was some danger (as measured by infertility and birth defects) to consanguineous marriages. Most states in the United States still forbid first-cousin marriages, and to this day, second-cousin marriages are permitted within the Catholic Church only by special dispensation. Some peoples, however, such as the Druze who favor uncle–niece marriages, still regard consanguinity as a way to strengthen family ties. Indeed, Christianity is the only major religion to proscribe cousin marriages. Until recently, the Chinese prized cousin marriage if the couple were the children of two brothers (geneticists call this parallel cousin marriage), but not if the unions were between offspring of a brother on the one hand and a sister on the other (cross-over cousin marriage). There is no genetic logic here, for in both cases, each marriage carries the same coefficient of relationship—husband and wife share an average one out of eight alleles (particular versions of genes) by descent. Of the roughly 100,000 genes they will each contribute to sperm or egg, about 12,500 will be the same alleles (the word refers to variants of any one gene).

The children of first-cousin marriages are more likely than those who are the offspring of unrelated persons to inherit two identical copies of the same gene. Products of even closer genetic unions, what we would call incestuous matings, share a much higher percentage of alleles. In the rare case of brother–sister unions, a child would have exactly the same two copies of a particular gene (be homozygous for them) in one-quarter of his or her total number of gene pairs. The few published studies of incest confirm that the offspring have a very high risk for being born with serious genetic diseases. Popular thinking has long held that marriage between relatives has the potential for producing children who are sublimely gifted or who are badly handicapped. In the case of Henri Marie-Raymond Toulouse-Lautrec Montfa, both predictions were realized.

The views held by geneticists about the impact of intermarriage on the health of offspring have vacillated significantly since 1900, and it is still difficult to estimate the magnitude of the risk (the theoretical benefits are now almost never discussed) associated with consanguinity. In the early 1900s, many scientists already believed that inbreeding conferred a very high risk of producing children who were mentally retarded, but there were few studies to support their opinion. The reasoning was that each of us must carry some alleles (gene variants) that in single doses pose no risk (and may even provide a benefit) but that if paired together (through inheritance from each parent) can have very harmful results.

Population geneticists estimate that our individual load of genetic lethals—gene variants that although harmless if only one copy is present either kill the individual or make it extremely unlikely that he or she will reproduce (infertility is genetic death) if two copies are present—is about 4–5. The reason that having one copy of any particular genetic lethal does not cause illness is simple. The protein made by the "normal" counterpart of the particular gene is sufficient to offset the lack of or malfunction of the other. In most cases that have been studied, it seems that people only need about 10% of the usual level of the normal protein to get by, so there is an adequate safety margin if one "bad" allele causes a 50% reduction in the output.

Children who are born of cousin marriages begin life with a set of genes that are drawn from parents with comparatively similar genomes (sets of genes). If a common ancestor of the cousins carried one of the harmful variants, it is statistically more likely that each cousin will have a copy than would be the case if they were unrelated. Therefore, the children of their union are more likely than the children of unrelated parents to inherit two copies of the "bad" gene. The very few studies of children who are born of highly incestuous matings, such as father–daughter rape, suggest that they are at much higher than average risk (perhaps 40%) of having a genetic disorder. This supports the argument for a genetic risk in cousin marriage, but says little about its magnitude.

Because humans have spent most of their history living in small groups, it is likely that matings between cousins were extremely common for eons. Our species tolerated this, but we have no idea at what cost. Almost certainly, there were many offspring of close genetic matings that did not survive. As recently as a century ago, until the rise of the automobile, first-cousin marriages were much more common in Europe and the United States than they are today. Studies from church records in Europe show that in the late 19th century, most people tended to choose spouses who had grown up within 15 miles of them, and many married neighbors within their small villages. Spouses were frequently second or third cousins, and not infrequently they were first cousins.

Throughout this century there has been a steady decline in first-cousin marriages, a trend that has made it possible to study their impact on the health of the offspring. Alan Bittles and James Neel, two leading population geneticists, recently reviewed the world literature on the effects of first-cousin marriage. They concluded that, contrary to earlier estimates, the average person carries only about 1 or 2 lethal genes that manifest in late pregnancy or childhood. If true, this suggests that first-cousin marriages in general pose relatively little danger for bringing out hidden genetic defects. However, any reassurance that one can draw about the risks of first-cousin marriage is of no solace to a particular couple who have a child burdened by a pair of genes that together cause severe disability.

Bittles and Neel could not rule out the possibility that there are many lethal alleles which kill the embryo so early that the pregnancy is not even recognized. It would be possible to investigate this by comparing the fertility (within the same culture) of cousin marriages with exogamous (outbred) marriages. A significantly lower fertility rate in cousin marriages would suggest the impact of early lethal genes. Another interesting conclusion that one can draw from their study is that the majority of changes (mutations) in human germ-line DNA are of no functional consequence.

TOULOUSE-LAUTREC: A SKETCH

The inbreeding in the Toulouse-Lautrec family was not at all atypical, especially for 19th century French nobility. The painter's grandmothers were sisters, and his mother's brother married his father's sister. Such unions, along with the law of primogeniture (that land passes to the first-born son), were essential for keeping together the vast estates that formed the basis of most great wealth.

Before he was three, it was apparent that Henri, the adored center of a powerful, doting, and, on his father's side, physically rugged, artistic, and profoundly eccentric, family, was not growing normally. He was not sickly, but he was frail. Contemporary photos, family letters, and the memories of those who knew him evoke an image of a tiny child with an unusually large head. At the age of six months he weighed only 10 pounds.

One of his biographers reports that Henri's mother was worried that his fontanel (the opening at the top of the head of newborns that normally closes during the first year) was still widely open when he was four, a sign suggesting a skeletal abnormality. Photos of him in early childhood show a large and unusually wide head. He spoke with a lisp, and his arms and legs were very short. Within his extended family Henri was 1 of 4 children from among 16 born in two marriages to suffer with a serious skeletal disorder that was almost certainly caused by having inherited two copies of a "bad" gene (a recessive disorder).

Throughout his childhood, Henri, who showed a penchant for drawing by the age of 3, grew slowly. At school, where he excelled in his studies, he was nicknamed "the little man." Fellow students, who loved the devil-may-care attitude with which he regarded his disability, used to carry Henri about so he would not exhaust or endanger himself chasing after them. As the years passed, he fell farther behind. At the age of 13 he was only 4 feet, 8 inches tall. Despite the boy's exuberant, charming personality, Henri's father, unable to control his disappointment that he had not sired a rugged hunter, distanced himself from his son, conduct that would continue through his life. His mother, devoted to Henri until his death, consulted doctor after doctor, none of whom offered more than reassurance and quack interventions, such as stretching him on a rack, to spur his growth.

On May 30, 1878, Henri, who at 13 was so frail that he walked with a cane, fell as he was rising from a low chair. The fall, which would not have harmed the average person, left him with a fractured left femur. Less than two years later, while walking in the garden, he slipped into a small ditch and fractured his right femur. With strict bed rest and splinting, the bones mended on both occasions. His growth, however, was over. He stood just shy of 4 feet, 11 inches tall.

As he moved through adolescence, Toulouse-Lautrec seemed to grow uglier. His nose broadened, his lips thickened, his legs bowed, and, some say that his hands grew much too large for his tiny arms. He nobly fought the disability. One summer in Nice he walked, swam, and rode for weeks on end. He swam he said, "like a toad, but fast and well." But still he could not take up his father's passions—riding and hunting. By the time he was 14, it was apparent that Henri would be burdened by a lifelong physical disability, one that was incompatible with the family traditions. Despite his mother's devotion, others began to push him outside the family circle. No doubt thinking that his son would never make a good marriage, Count Alphonse early on took the decisive step of writing a will in which he left his vast tracts of land in southern France to Henri's younger sister, Alix.

It was about then that Henri began to disappear into his art. In 1880 when he was 15, he made 300 drawings and about 50 paintings. In the autumn of that year he and his mother (who was estranged from his father) moved to Paris. Henri was to sit for his baccalaureate exams in the spring of 1881, in that era typically the end point for the education of a son of a wealthy family. But soon after arriving Henri began to spend his days in the studio of a family friend, Rene Princeteau, a connection which his father had made for him. The solid artistic reputation enjoyed by Princeteau, who was then 37 years old, was especially remarkable because he had been born deaf. The degree to which he had triumphed over his disability impressed everyone. He could read lips well, and with a bit of practice one could understand his speech. Throughout Henri's first years in Paris, Princeteau treated him like a son and by example showed him what he might become. Henri quickly abandoned his formal schooling to pursue painting, a course that caused him to fail his first try at the baccalaureate. To please his mother, he did manage to buckle down enough to pass on his second attempt.

Henri's presence in the art world grew quickly. Within a year, at Princeteau's urging, he had enrolled in the studio of Leon Bonnat, a member of the French Academy of Beaux Art and a leading teacher. Along with his fellow young artists, Henri soon immersed himself in cafe life, started a lifelong pattern of heavy drinking, and became an afficionado of the darker sides of Paris. More and more he was drawn to the bohemian section of the city known as Montmartre.

In the late 1880s, Henri turned to lithography. His first poster, The Moulin Rouge, appeared in 1891 and, although (or perhaps because) it broke sharply from the dominant style, was immediately hailed as a tremendous success. His bold compositions in which the subject dominates the foreground, usually in three-quarter profile, ruled the French poster-making world for more than a decade. Not burdened by convention, Toulouse-Lautrec was willing to create on commission for virtually anyone who wanted his services. During the 1890s he illustrated song books for composers and menus for restaurateurs, designed posters for books and journals, crafted billboards for theatre productions, and sketched advertisements for furniture stores and bicycle factories. His work was decidedly antiacademic, and the Parisians loved it.

As he grew older and realized that hopes of normal relationships with women were remote, he took his solace more and more in alcohol and the brothels that dotted that quarter of Paris. He befriended many prostitutes, spent lavishly on them, and on occasion disappeared into bordellos for weeks on end. He is, for example, known to have lived for months at a time in an upscale bordello at 5, rue des Moulins, near the Opera, a luxurious establishment famous for having rooms decorated in different historical periods, apparently to satisfy the fantasies of some of the regular clients. Affectionately known as "Monsieur Henri," he took his meals with the women when they were off duty, and painted many of them and the lavish rooms in which they plied their trade. His scenes of those interiors, of the madams, the laundrymen, the prostitutes submitting to medical exams or sitting idly while waiting for their next clients, are among the most important artistic records of that life. He painted many of the women, and it is fair speculation that he tossed off hundreds of sketches in those times that never saw the light of day. On one occasion toward the end of a life by then dominated by drinking, he moved his studio and left 87 works behind for the landlord, saying, "They are of no importance."

Even though he had abused alcohol for two decades, Toulouse-Lautrec remained functional until his mid-30s. He did suffer from severe dental problems, and he broke one more bone, his clavicle, during a drunken fall downstairs. He also almost certainly contracted syphilis and may have suffered from neurosyphilis. On one occasion his alcohol abuse became so severe that his friends and family had him committed to an asylum in a last-ditch effort to save his life. After a bout of delirium tremens, he emerged improved, only to relapse quickly.

In 1901 Lautrec, his health worn down by two decades of alcohol abuse and the insidious march of syphilis, left Paris to spend the winter in Bordeaux with his friend, Paul Viaud, whom his mother had hired to stop him from drinking. His art works from this period seem all too obviously to reflect a deep depression. The clear bright colors of the famous posters have given way to gray, dark, heavy sketches, images that suggest despair and death. In August, while staying in a small coastal town, Lautrec suffered a stroke. He was taken to his mother's estate at Malrome, where he fell into a coma and died on September 9. As she had been figuratively his whole life, his mother was at his side. His father too was nearby. According to one story, Henri's last words were to call him an "old fool."

Despite his disability (or perhaps because of it) and dissolute life, Henri left behind hundreds of oil paintings and the posters that soon garnered him fame, as well as thousands of sketches. His father rejected any suggestion that his son's work was important and enduring and signed away his paternal rights to them. Today, two of the paintings that he rejected grace the entrance to the top floor at the Musee D'Orsay in Paris, which houses the greatest collection of Postimpressionist French painting in the world. His mother, who lived until 1930, worked devotedly to create the Musée Toulouse-Lautrec in Albi, France, where today many of his works hang.

PYCNODYSOSTOSIS

There are many genetic causes of dwarfism—a term used originally to describe a person with short stature whose limbs and torso are not proportionate. Until recently, the main tools used by clinical geneticists to categorize and study affected persons were radiographs and close attention to clinical features. Some conditions, such as diastrophic dwarfism, are often fatal. An affected infant may die of respiratory failure if the rib cage is too weak to help the lungs to expand properly. Other forms, such as the relatively common and well-recognized achondroplasia (a diagnosis that one early medical detective erroneously applied to Toulouse-Lautrec), are compatible with a long, healthy, productive life, limited all too often only by the prejudice of strangers.

Periodically for the last 80 years physicians have speculated about the cause of Henri's disabilities. Most of the guesses—achondroplasia, osteogenesis imperfecta (the bones of affected persons break almost without being stressed), and epiphyseal dysplasia (a disorder of the growth plates of the long bones)—were not strongly supported. In 1962 Maroteaux and Lamy, two prominent French doctors, described a form of dwarfism that they called pycnodysostosis (the term means dense bones), and then garnered more than the usual attention by claiming that Toulouse-Lautrec, by then a star in the French artistic pantheon, had been afflicted with the condition. The records of Henri's physical condition seemed to match with the features of their living patients. In addition to a tendency to fracture their leg bones, persons with pycnodysostosis have small stature, a large head with a prominent forehead, and dental problems. Although many doctors doubted the posthumous diagnosis, it stuck, and Toulouse-Lautrec was inducted into a second select group—famous people who have been given genetic diagnoses long after death. He was sometimes mentioned in genetics courses as an example of the dangers of consanguinity. Professors speculated that each of his parents had contributed a copy of the allele which causes pycnodysostosis or some other allele which affects the growth and strength of long bones.

In 1995 a group of geneticists, including Bruce Gelb and Bob Desnick, who work at Mount Sinai School of Medicine in New York, rekindled interest in pycnodysostosis when, using new DNA mapping techniques in an Arab family with many affected members and in which there had been many consanguineous marriages, they showed that the culprit gene was located somewhere within a small region on the long arm of chromosome 1. Their scientific report on the mapping of the pycnodysostosis gene was accompanied by an essay by Julia Frey, an art historian who wrote a fine biography of Toulouse-Lautrec. Although she interpreted the historical data to cast doubt on the diagnosis, on balance I think the material supports it.

In 1996 the New York team successfully cloned the gene responsible for pycnodysostosis, showing that the disorder is caused by defects in a gene that makes a protein called cathepsin K. This protein is used by cells called osteoclasts that are primarily responsible for remodeling bone, especially during growth. The osteoclasts of people with this very rare disease are able to perform part of the remodeling process, but because their cathepsin K does not work properly, the cells are unable to finish their job. Because pycnodysostosis is essentially a disease that only affects osteoclasts, it may turn out to be treatable, either by transplanting normal osteoclasts or through gene therapy (see Chapter 20).

Can we be sure that Toulouse-Lautrec suffered from pycnodysostosis? No. Could the diagnosis be made definitively or excluded? Yes. There are two possibilities. If descendants of his sister (who, assuming the diagnosis is correct, would have had a 2 in 3 chance of having inherited one copy of the allele) provided blood samples in which DNA analysis showed the presence of a copy of a mutated cathepsin K gene, it would by inference clinch the historical diagnosis. If descendants were to successfully petition a court for permission to exhume the artist's remains, then DNA analysis of a bone fragment might provide a definitive answer. Dr. Robert Desnick, the physician-researcher who led the team that worked out the molecular basis of pycnodysostosis, thinks that Toulouse-Lautrec was affected with it. He has discussed the possibility of resolving the diagnostic question with the family, but the descendants were reluctant to be tested and have ruled out exhumation of the artist's remains.

Those geneticists who are fond of offering Henri Toulouse-Lautrec as an example of inbreeding gone awry rarely mention the many talented artists among his ancestors. Julia Frey wrote that, "For three generations Henri's forebears had shown a marked artistic talent. His great-grandfather, Jean-Joseph de Toulouse-Lautrec, had done a number of small, but excellent, family portraits." Naturally, since he was a nobleman, that talent served him only as a hobby. It would have been unthinkable for him to become a professional artist. According to family letters, two of Henri's uncles were also gifted amateur painters. There is, of course, no single gene that governs whether one is capable of becoming a talented painter. But Henri was drawing well without benefit of a tutor by age 3. As an adult, his art is all the more impressive given the deformity of his limbs, which surely must have affected his brush strokes. He actually had to use unusually long brushes so he would not have to sit too close to the canvas. Nevertheless, his paintings and sketches are wonderfully vibrant.

Given the family history, it is hard to avoid wondering whether there was also a genetic basis for the Toulouse-Lautrec line's legendary history of alcohol abuse. Fantastic stories of their love of and capacity for drink stretch back through the centuries, but Henri's drinking seemed to eclipse all the ancestral tales. Did his parents' union provide a genetic basis for that predisposition? We will never know. The many well-documented stories of his behavior also suggest that Henri's father suffered from manic-depressive illness. If so, Henri had a fairly good chance (family studies put the risk to sons of affected parents at about 10%) of himself suffering from this disorder, an illness that drives many persons to alcohol abuse. Severe alcohol abuse is encountered more often among those with bipolar disorder than among persons who suffer from depression. Perhaps the

Toulouse-Lautrecs were driven to drink in part to cope with the genetic burden of madness. If so, Henri's alcoholism was not necessarily a response to having inherited the two defective cathepsin K genes that destined him to suffer from pycnodysostosis.

Today, through advances in physical therapy and orthopedic surgery, persons with pycnodysostosis do moderately well. But it seems highly unlikely that another affected person will ever live so on the edge or paint with such fervor or manage to capture the spirit of Paris so well as did Toulouse-Lautrec at the Moulin Rouge. In a span lasting less than 20 years and hampered by physical deformity and severe alcohol abuse, he produced 737 canvases, 275 watercolors, 368 prints and posters, and 5,084 drawings. Almost certainly, he created hundreds of other works that he destroyed or simply left behind in the brothels and dance halls of late 19th century Paris. As one turns the pages of an art book or strolls through the museum at Albi, it is hard to think of Toulouse-Lautrec as a man with a disability.

Czar Nicholas II and his family. *(© Illustrated London News/Archive Photos/Picture quest.)*

Old Bones
DNA and Skeletons

THE ROMANOVS

On June 5, 1995, Pavel Ivanov, a 42-year-old molecular geneticist who was working at the Engelhardt Institute of Molecular Biology in Moscow, arrived at the United States Armed Forces Institute of Pathology in Rockville, Maryland, with an unusual package: a piece of the femur and a piece of the tibia of Grand Duke Georgij, the younger brother of Imperial Russia's last tsar, Nicholas II. Eleven months earlier, under the watchful eyes of bishops of the Russian Orthodox Church, forensic scientists had removed some of the Grand Duke's remains (he died of tuberculosis at the age of 28 in 1899) from his Italian marble coffin in the Cathedral of St. Peter and St. Paul in Petersburg. Ivanov also carried two other curious and precious items: a slender section of a handkerchief from a museum in Japan that is reported to be stained with the blood of Nicholas II, and a strand of hair from a locket that for decades reposed in a St. Petersburg palace and is said to have been cut from Nicholas II when he was a child. The Russian scientist hoped that DNA analysis of these artifacts would close one of the most persistent questions in modern Russian history: What really happened to Nicholas II and his family?

The Russian Orthodox Church had permitted the exhumation because it needed irrefutable evidence to corroborate the dramatic reports that the skeletal remains of the last Romanov tsar and his family had been identified. If DNA analysis of Grand Duke Georgij's mitochondrial DNA (see below), which would be extracted from bone samples for which the authenticity was beyond doubt, indicated a match with the mitochondrial DNA extracted from the bones alleged to be those of Nicholas II, it would establish to a certainty that the nine fragmented skeletons that Geli Ryabov, a filmmaker, and Alexander Avdonin, a geologist, had located in a shallow grave in Yekaterinburg in the Ural mountains included five members of the imperial family, all of whom had long been believed to have been assassinated in compliance with an order issued by Lenin in July, 1917.

According to the official report filed in Moscow in 1917 by Yakov Yurovsky, the man who carried out the execution, his squad had killed eleven people: the tsar, his wife, their five children, three servants, and the family physician. The discrepancy between that number and the nine skeletons found in 1995 could be explained by the fact that Yurovsky reported that he had incinerated two of the bodies before deciding to bury the rest. Of course, the discrepancy in the number of skeletons also indirectly supported the long popular rumor that Anastasia, one of the tsar's daughters, had miraculously escaped the death squad.

Confirmation of the identity of the skeletons would be an important historic and political event. Among other things, it would provide the historical evidence needed by the Russian Orthodox Church to reinter the last of the ruling Romanovs, this time with full religious rites. A mass funeral for the tsar's family would rivet the nation and further strengthen the position of the Orthodox Church in post-Communist Russia.

By the summer of 1995, efforts to confirm the identity of the Yekaterinburg massacre—by merely looking at the skulls and bones, one could see evidence of crushing blows, bayonet thrusts, and bullet wounds—had become mired in politics. The grave had been rediscovered in 1979, but it was only in 1991 after the collapse of the Soviet Union that the government permitted the exhumation, which was carried out the day after Yeltsin's inauguration. For a full year, Dr. Sergei Abramov, a forensic scientist and bone expert in Moscow, worked to reassemble 700 bones and several hundred fragments into nine skeletons. After painstaking analysis, he was convinced that the remains were of Tsar Nicholas II, his family, and servants, but he realized that the only way to close this extraordinary moment in 20th century history was to have other experts confirm his conclusions.

In the summer of 1992, Dr. William Maples, a forensic pathologist from the University of Florida, and fellow scientists who are experts in dental identification and hair analysis went to Yekaterinburg at the invitation of local authorities to study the skeletons. Just a few weeks earlier, acting on behalf of the Russian Ministry of Health in Moscow, Dr. Ivanov had reached an agreement with the forensic section of the British Home Office to have one of their scientists, Dr. Peter Gill, undertake two tasks: determine whether the skeletons were part of a family group, and, if they were, determine whether DNA analysis of the remains matched that of DNA from living descendants of the imperial family. Prince Philip, Duke of Edinburgh, consort to Queen Elizabeth II, is the grandnephew of Empress Alexandra, the tsar's wife. Because an unbroken female line connects Prince Philip to Alexandra and her children, his DNA could be used to confirm the identity of the skeletons. He agreed to provide a blood sample.

Efforts to extract DNA from old bones and use it to establish identity began in 1984 when Russell Higuchi, one of the scientists who developed PCR (polymerase chain reaction, a technology that enables one to amplify large amounts of DNA from exquisitely small initial samples), successfully retrieved mitochondrial DNA sequences from the remains of a 140-year-old quagga, a recently extinct species that looked like a cross between a horse and a zebra. A year later, an Italian scientist reported cloning DNA from an Egyptian mummy, and the race was on. One spectacular event was the successful cloning of human DNA from 91 well-preserved brains found among the 177 skeletons buried about 7,000 years ago in a heavily silted lake in Florida. That site is thought to have been used by a primitive culture for ritual sacrifice. The remains were well-preserved, in part because of the low oxygen conditions in the lake bed. By 1992, forensic scientists and molecular anthropologists had sufficient experience with ancient DNA that they were eager for almost any challenge, especially if large bone samples were available.

From September 1992 until early 1994, Dr. Gill and his team labored to extract DNA from fragments of the femurs of each of the skeletons and, by comparing the sequences of the DNA letters, use it to determine which, if any, of the skeletons belonged to the same family. Using the small amount of nuclear DNA he was able to extract, Gill proved that one of the skeletons of the adult males in the group had to be the remains of the father of the three young women whose remains were also in this grisly collection. This fit the profile of the Romanov family, but there were still several issues to be resolved, and there was not enough nuclear DNA to complete the molecular analysis.

The group turned to mitochondrial DNA. Mitochondria are tiny structures that reside in the cytoplasm that surrounds the cell's nucleus. They each contain a small circular genome (16,569 bases) that codes for key genes involved in making proteins that are critically important for operating the cell's power plant. Evolutionary biologists speculate that mitochondria are the ancient remnants of a subcellular organism or a simple bacterium that colonized higher organisms hundreds of millions of years ago, and that over the eons made itself indispensable. Only over the last decade or so have we come to realize that there is a whole class of genetic diseases that are caused by mutations in the mitochondrial genes and that are not inherited according to Mendelian laws.

Because it is not found in sperm, mitochondrial DNA is inherited strictly matrilineally. It has been transmitted from mother to offspring throughout the history of our species. Thus, mitochondrial DNA provides scientists with an unusually powerful tool to study the origin and spread of the human family. By comparing differences in the sequence of mitochondrial DNA, scientists can estimate when different human groups diverged. The greater the differences, the more ancient the divergence. In effect, mitochondrial DNA constitutes a kind of biological record of the human diaspora.

One recent scientific report was popularly interpreted to show that all humans could theoretically trace their ancestry to a common maternal relative who lived in Africa about 200,000 years ago. The journalistic announcement that there really was a human Eve grew out of study of seven regions of the mitochondrial DNA taken from 147 persons who had been born in five distinct geographic regions of the world. The scientists used the variations in the patterns among the different persons to construct a phylogenetic tree, a diagram that defines degrees of relatedness according to the amount of divergence (essentially, the accumulation of differing mutations over time) between the mitochondrial DNA sequences. As one moves down the tree, its branches converge on a trunk—one ancient mitochondrial DNA pattern, the pattern from which all others diverged. The woman from whom this mitochondrial pattern was obtained is, arguably, the oldest true female ancestor to whom we are all related.

The Russian skeletons yielded abundant mitochondrial DNA. Analysis of the samples from the skeletons presumed to be Empress Alexandra and three of her four daughters perfectly matched the mitochondrial sequence of the sample donated by Prince Philip. The next task was to find living relatives of Nicholas II so the forensic scientists could attempt to establish his identity. This was not so easy, but they eventually located two living relatives, Xenia Cheremeteff-Sfiri, a direct female descendant of the tsar's sister, and James Carnegie, the Third Duke of Fife in Scotland, who descends from a line that is connected to Nicholas II through his grandmother. Both agreed to donate blood for analysis. The mitochondrial DNA samples taken from these two matched each other perfectly but, to Gill's surprise, they did not match with the sample from what was presumed to be Nicholas II. When he sequenced the DNA in the putative tsar's mitochondrial DNA, Dr. Gill found a mixture of the DNA letters C and T at position 16,169. In the mitochondrial DNA taken from Xenia Cheremeteff-Sfiri and James Carnegie, the DNA at position 16,169 is always a T. There can be more than one type of mitochondrial DNA in the cytoplasm of a single cell because hundreds of these subcellular structures are transmitted via the human egg through the generations, and mutations can arise in any one of them.

Probing further, Gill and his team found that Nicholas II had two forms of mitochondrial DNA (a condition called heteroplasmy), which is not uncommon, but which in this instance cast a tiny shadow of doubt on the argument that the remains were those of the tsar. Although the scientists continued to seek other ways to establish identity, they were ready to make their case. In July, 1995, they announced that their DNA studies established at a level of certainty greater than 98.5% that the bones were of the tsar and his family. Dr. William Maples and his colleagues were quick to challenge the assertion, arguing that the results of the DNA studies done on the remains could have arisen if two samples had been contaminated. This made officials in the Russian Orthodox Church unwilling to accept the evidence as definitive, and led them to acquiesce to the exhumation of Nicholas's brother, Georgij.

In the fall of 1995, another round of DNA tests confirmed that the remains presumed to be Nicholas II were in fact those of a brother to Georgij. Ivanov and colleagues at the DNA identification laboratory operated by the U.S. Armed Forces Institute of Pathology showed that the Grand Duke's mitochondrial DNA matched that of the putative tsar's exactly, even to the point of having a mixture of C and T at position 16,169. Thus, the heteroplasmy that thwarted the first effort to prove identity became the crucial proof in the second attempt. The reason that Countess

Xenia's mitochondrial DNA did not match at that position is easily explained by a "bottleneck" hypothesis. Simply put, only some of the many mitochondria in a human egg are transmitted to the next generation. If there are two different mitochondrial populations (as defined by slight differences in DNA sequence), then chance alone can account for one relative having one lineage while another did not.

One part of the story, the best known and most popular, still needed resolution. Until she died in a Virginia nursing home in 1984, a woman named Anna Anderson claimed she was in fact Anastasia, Nicholas II's youngest daughter, who had escaped the firing squad's bullets. In 1979, Anna Anderson underwent surgery, and a bit of her intestine was stored in a pathology department in a Virginia hospital. Using that sample, Dr. Gill's team showed absolutely that she was not a daughter of the tsar, but that she was related to the maternal grandnephew of a Franziska Schangkowska. The woman who had spent her life claiming she was the tsar's daughter was, as had long been thought, actually born into a Polish working-class family.

Perhaps she would have been happy to learn that we probably all number a royal ancestor or two in our family tree—if we go back far enough to look. Consider. We have two parents, four grandparents, eight great-grandparents and sixteen great-great grandparents. If we go back just 20 generations, about 400 years, and if we assume no inbreeding, we can theoretically claim relatedness with 2 to the 20th power ancestors, some 1,048,576 persons, one of whom might well have been part of a royal family ... or at least well off.

Proceed back just 10 generations more, another 200–250 years, and one could theoretically claim 1,000,000,000 ancestors. But this number far exceeds the total number of people who then lived on the planet. How can we reconcile the theoretical prediction and the reality? In fact, one need go back little more than about 10 generations before discovering that one is related distantly to all sorts of people including, perhaps, the very neighbors one most detests. Like it or not, we really are a human family. We share our ancestors. Go back far enough and you will find many great-great-grandparents to whom you are related through multiple other relatives. If the technologies for extracting DNA from old bones and studying it to determine relatedness improve much more, it will be possible to connect almost any recent human fossil (say within 5000 years) we find to a living descendant. Indeed, we are nearly there.

CHEDDAR MAN

In 1903, cavers exploring Cheddar Gorge in southwest England discovered a remarkably intact skeleton of a man that, on the basis of archaeological study of artifacts in the cave, appeared to have lived more than 5000 years ago. Later on, carbon dating, a measure of radioactive decay in the bones, pinpointed his death at about 7150 B.C. In 1995, a British newspaper intrigued by the work of Bryan Sykes, an Oxford molecular biologist who is an expert on ancient DNA and who was among the first to show that it could be obtained from old bones, asked him to try to find a living descendant of Cheddar Man, as the fossil is fondly called. Sykes, who had just performed the DNA studies of the 4000-year-old Ice Man discovered in a glacier on the Italian-Austrian border in the summer of 1995, took the challenge.

Typically, the curators were stingy. They permitted Sykes to remove a single tooth from the fossil jaw. Using a drill, he and his team retrieved a core sample from the region of the tooth known as the dentine. They found a surprisingly large amount of DNA, which after weeks of laborious study they were able to characterize with confidence. Having constructed the mitochondrial DNA profile of the fossil, they were ready to look for relatives. The search was based on some simple logic. They called for volunteers who to the best of their knowledge came from families that had lived in or near Cheddar for generations. After interviews, the team invited a mere 20 persons to give blood. To his immense surprise, Eldrian Targett, a local history teacher who lives just a mile from the caves, learned that he and the 9000-year-old fossil were both descendants of a common female relative. Targett can, for the moment at least, claim an ancestor older even than can members of the world's most studied royal families. When she learned that her husband was related to a caveman, Targett's wife's quipped, "Maybe this explains why he likes his steaks rare."

The Cheddar caves are owned by the Lord of Bath, one of England's richest men. Intrigued by the mitochondrial studies, the Lord asked Sykes to analyze his DNA as well. Over dinner in a Victorian era hotel, Professor Sykes, a blonde, ruddy-cheeked, puckish fellow, confided to me that the Lord's butler was more closely related to the ancient skeleton than was his master, who, one might surmise, had been eager to establish that he had a more ancient lineage than any other member of the House of Lords.

Sykes's success in establishing a family connection between a living Englishman and a 9000-year-old skeleton set off a media frenzy. The

British tabloids gave new dimensions to the definition of bad taste. One ran the story below photos of two topless models holding fake stone age axes. Even the normally staid science journals delighted in speculating how far back in time one could retrieve stretches of DNA that were sufficiently intact so that questions could be asked about the organisms from which they were derived. The work with Cheddar man fueled the fantasies that had been evoked by *Jurassic Park*.

The idea for that novel arose when Michael Crichton read reports that molecular biologists had successfully extracted DNA from the fossilized digestive tracts of insects that had been preserved intact for tens of millions of years in the hardened amber resin in which they had become stuck and perished. It is logical to suppose that the ancient mosquitoes fed on dinosaurs, so it is conceivable that their bellies might include sequences of dinosaur DNA. It is, however, an amazingly long leap to think that one could reconstruct a whole genome of an extinct creature by assembling hundreds of thousands of tiny fragments. It is next to impossible that all the sequence is present, even in tiny fragments, and even if it were, the task of assembling it is beyond current technology.

The reports of extracting DNA from the remains of organisms that died many millions of years ago may not even be accurate. Sykes is among those who dismiss most of the claims out of hand, attributing the reports at best to sloppy science and at worst to fraud. He points out that the laws of physical chemistry that govern the effects of the impact of air and water molecules (oxidation and hydrolysis) on DNA suggest an outer limit of about 100,000 years for the molecule to remain intact, unless, of course, it has been perfectly protected as in amber. In 1997, a group of British scientists reported that despite meticulous efforts they had been unable to obtain any DNA from a large series of amber specimens. Their report contrasted sharply with 13 of 14 earlier, but much smaller, efforts that had claimed success. Reviewing the earlier papers, Sykes concluded that the DNA that each researcher claimed to have found was actually a modern contaminant. In one case, it was almost certainly a fragment of human mitochondrial DNA. In another study that claimed to detect DNA from a 20-million-year-old fossilized magnolia leaf, Sykes argues that the DNA is from a modern magnolia. It appears that we are unlikely ever to retrieve and study long stretches of DNA from specimens much older than 100,000 years.

Yet, this is still a large enough historical window to let us explore an endless list of fascinating questions about our own past. For example, mitochondrial DNA is already being used to investigate much more important historical questions than whether the Lord of Bath is a member of the Cheddar cave family. In 1995, archaeologists in Honduras uncovered a royal tomb that they think contains the remains of Kinich Ah Pop, a 5th century ruler in the Mayan Copan dynasty. There is strong circumstantial evidence that the remains are those of Kinich, but to confirm his royal lineage, scientists will compare his DNA to that taken from skeletons for which the evidence of membership in Mayan aristocracy is even greater. In Spain art historians and geneticists hope to celebrate the 400th anniversary of the birth of the great artist, Diego Velazquez de Silva, by locating his tomb and reinterring his bones. Since early 1999 archaeologists have been digging at the Plaza de Ramales, the former site of the Church of San Juan where Velazquez was buried. They hope to find the remains of his tomb and verify the bones by comparing the DNA from bone fragments with that of known living descendants. Similar work is surely to be done in archaeological projects around the world for many years.

In 1999, Dr. David Goldstein, a geneticist at Oxford University who is attempting to develop genetic signatures of scattered Jewish populations in order to refine our understanding of the diaspora, reported a remarkable discovery. In southern Africa the Lemba, a black Bantu-speaking tribe, have an oral history that they were led out of Judea by a man named Buba. They practice many Jewish customs such as circumcision, and they shun pork and the flesh of any pig-like animal. Goldstein and his colleagues demonstrated that 9% of Lemba men have a DNA sequence on their Y chromosomes that is distinctive of the cohanim (the Jewish priests who are believed to be the descendants of Aaron, the brother of Moses). Interestingly, among those Lemba men in the most senior of their twelve groups, 53% have the cohanim DNA markers. The genetic evidence combined with the cultural history of the Lemba strongly support their claim to Jewish ancestry.

DNA analysis of old bones can even be used to establish cause of death, a form of sleuthing that can provide insight into historical documents. In Luke 4:27 it is written that while in the Jordan Valley, Christ cured a leper named Naaman the Syrian. In 1994, archaeologists uncovered bones from the ancient monastery of St. John the Baptist in a valley near Jericho. Carbon dating indicated that the bones are 1400 years old. Working at University College London, microbiologist John Stanford and surgeon Mark Spigelman extracted DNA from the dried-up marrow of a toe bone. Using DNA probes encoding sequences from *Mycobacterium leprae,* the bacterium that causes leprosy, the scientists demonstrated that the bone came from a man who had been afflicted with the disease. This effectively proves the presence of this disease in the Jordan Valley at about 600 A.D.

Spigelman argues that this bacterial sleuthing can answer other very important questions. For example, did syphilis originate in the Old World or the New World? For years, historians have asserted that the sailors on Columbus's ships were the vectors who carried the disease to the Americas. If DNA from the spirochete (the bug that causes syphilis has a corkscrew or spiral shape) could be detected in a New World skeleton that was unquestionably older than the 15th century, it would refute that claim.

Of far greater importance is the current effort to use molecular analysis to decipher the genetic code of the RNA virus (RNA is a molecule very similar to DNA that takes its molecular orders to the cytoplasm where it directs the assembly of proteins) that caused the 1918 flu pandemic which killed more than 600,000 Americans and as many as 25,000,000 persons around the globe. Because RNA viruses are extremely fragile when outside of their host cells, there are few places on earth where one might find an intact copy of the agent that caused death in 1918. One is some graves near the coal mining village of Longyearbyen in Norway.

In the winter of 1918, seven healthy young miners working there caught the flu, succumbed in a matter of days, and were promptly buried in the permafrost. Legend has it that dynamite was used to prepare their graves. Thinking that they might be able to track down intact particles of the flu virus, an international scientific team was recently granted permission by the Norwegian government to exhume one or more of the bodies and take tissue samples in which they will search for dormant virus. The hope is that if they find a complete flu genome, the scientists could infer why the virus was so deadly and set the foundation for developing a vaccine to protect the world from an attack by a similar agent in the future. One of the issues that Norway's public health authorities faced was whether there would be a danger of awakening a dormant virus of great lethality. Almost all virologists agreed that this is extremely unlikely, but few were willing to say it is impossible. In the end, the remote risk was viewed as far smaller than the potential benefits that might flow from the research.

Identity testing offers unending adventures to the DNA detectives. James Starrs, the forensic scientist who testified at the Baltimore trial concerning the remains of John Wilkes Booth, also turned up at an interesting trial in Missouri. History recounts that Jesse James, one of America's most notorious outlaws, died on April 3, 1882, when Bob Ford, a 21-year-old who had just joined the gang, killed him for the $20,000 bounty. For decades a few amateur historians have argued that James was too clever to be killed by Ford, and that the murder was a setup to cover James's getaway and retirement from outlaw ways. In 1995, the Missouri judge granted permission to the descendants of Jesse James to open his grave. On their behalf, Starrs compared DNA from a bone and from a lock of hair taken from the grave with DNA from two of James's descendants. Unfortunately, for those who would prefer a more romantic tale, the DNA tests showed that the skeleton is that of an ancestor of the living relatives, and thus must be the remains of the celebrated outlaw.

Ironically, other forces may be foreclosing the opportunity for anthropologists, archaeologists, and molecular biologists to use DNA to solve important questions. Just as we have acquired a greatly enhanced ability to study ancient DNA, there has emerged a strong trend to return remains that have been housed in museums for a century or more to native peoples for reburial. The Museum of Victoria in Australia recently gave back to aboriginal peoples remains that may be 15,000 years old. In 1991 the Smithsonian Institute returned 756 sets of skeletal remains and funerary objects to a tribe that lives on Kodiak Island in Alaska. In Israel a new law requires the reburial of all skeletons less than 5000 years old, a rule that some scientists argue ends anthropology in that nation. The new laws make no allowance for retaining a small bit of bone for DNA studies.

The decision to return remains is motivated in part by guilt over the many atrocities committed against indigenous peoples during the colonial era, a time when simplistic racial thinking was at its zenith. For example, the Australian Aborigines, whom 19th century anthropologists thought represented our missing link to the great apes, suffered untold insults ranging from grave robbing to murder as a means to collect specimens. Another force is surely our great sensitivity to the need to act as stewards of the planet and respect biological diversity in all forms. However, when remains are so old that there is no possible cultural connection with living persons, it seems fair to decide the fate of old bones by asking what use of them best serves the human family. The answer to that may sometimes be that greater good comes from studying them before reburying them.

In the summer of 1999, hikers crossing a glacier in western Canada discovered the remarkably intact remains of a man who almost certainly died before modern Europeans reached that region. If the body turns out to be extremely old (say several thousand years) DNA studies might shed light on the migration patterns from Asia to North America. In planning their studies of the body, scientists are working with representatives of two indigenous groups, both of whom assume the body may be that of an ancestor. In sharp contrast to other, similar situations, because of the respect shown by the scientists to the beliefs of the native Americans, the two tribes have cooperated with the research. It is hoped that this will provide a model for research with other discoveries.

JUSTICE

The DNA Revolution in the Courts

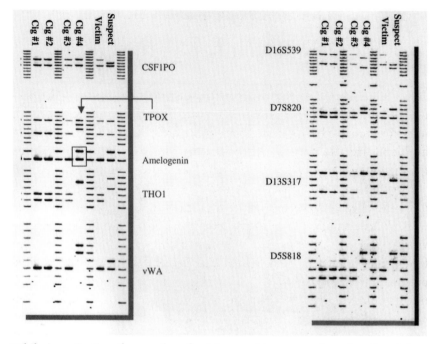

While investigating the murder of a 36-year-old woman, detectives found four cigarette butts. DNA analysis shows that cigarettes 1, 2, and 3 were smoked by the victim. The box around the Amelogenin marker shows that cigarette 4 was smoked by a man. However, the identity profile of the DNA extracted from cigarette 4 does not match that of the original suspect (shown here). Months later the profile of the DNA on cigarette 4 was shown to match that of another suspect. Thus, this DNA evidence was used both to exclude an innocent suspect and to support the prosecution of the perpetrator. (Courtesy of Cecelia A. Crouse, PhD., Supervisor, Serology/DNA Section, Palm Beach County Sheriff's Office Crime Laboratory, West Palm Beach, Florida.)

DNA Detectives
The New DNA Evidence

Snowball, a white American shorthaired cat who now lives with the parents of his original owner on Prince Edward Island, may be the first feline in history to have been served with a court order demanding a sample of his blood. He also must be the first cat that has ever provided evidence leading to the conviction of his master of second-degree murder.

In October, 1994, Shirley A. Duguay, a 32-year-old mother of five children, disappeared from her home in Sunnyside, a city of 16,000. Three weeks later, a military team out on maneuvers about five miles from the victim's home discovered a plastic bag containing a man's bloodstained leather jacket. At the scene detectives found several white hairs, which they presumed to be human, in the coat lining, a discovery that they thought might help to identify its owner. The hairs turned out to be from a cat. The police already knew that Douglas Beamish, Shirley's boyfriend and the father of three of her children, owned a white cat. Based on circumstantial evidence, they strongly suggested that Beamish had killed Shirley Duguay, but without a body no arrest was made. The search for the body, which was interrupted by a harsh winter, ended on May 6, 1995, when Ms. Duguay's remains were unearthed from a shallow grave. Soon thereafter police arrested Beamish, who was charged with murder.

When police inspector, Roger Savoie, learned that the white hairs were from a cat, he assumed that he could simply obtain a blood sample from Beamish's cat, Snowball, and compare the DNA profile of the crime scene hairs with the profile of the cat's blood. But the first several forensic DNA labs that he called were reluctant to take on the job. They had no experience with analyzing cat DNA and no feline reference population to provide the data with which to calculate the odds that a match between the hairs on the jacket and Snowball's blood could occur by chance alone. Making call after call, inspector Savoie eventually reached Dr. Stephen J.

O'Brien, Chief of the Laboratory of Genomic Diversity at the National Cancer Institute in Frederick, Maryland. Dr. O'Brien, who leads a team that is developing a linkage map of cat genes (organizing their linear order on the cat chromosomes), had just published a paper characterizing the location of 400 DNA markers known as short tandem repeats (STRs) on the cat genome.

An STR is simply a short stretch of repeated units of DNA letters. For example, CACACACACA is a five-dinucleotide CA repeat (the C and A are the symbols for two of the four chemical building blocks of DNA). In cats, humans, and virtually every other higher organism, there are many stretches of road on the vast DNA highway that do not code for proteins (some people call these noncoding regions "junk" DNA), but which contain STRs. STR loci are highly variable. At a locus where you might have 9 CAs in a row, your next-door neighbor might have 12, and I might have 7. Because the technology to demonstrate the difference between samples due to variation in the length of the STR is well-established, this diversity offers a way to determine whether or not two samples derive from the same individual. Because there are many STR loci at different addresses on the chromosomes, it is possible to use a consensus set of them to determine whether two DNA samples come from the same individual. Much work has been done in assembling a set of STR loci to use in matching human samples. Except in the case of identical twins, the odds of two randomly selected persons having the same flavor of STR at just 5 loci is vanishingly small. In general, if two samples match at even 3 of the 13 loci currently used in human identification studies, they are highly likely to match at as many more as are tested. Conversely, if two samples do not show exactly the same STR number at even 1 locus, it is almost certain that they are not from the same individual.

The harder question is to decide how many exact matches one must demonstrate to argue convincingly that the samples are identical. During the years from 1987 to 1996, this was the central question of DNA forensics. The debate heated up in 1992 when a Committee on DNA Forensic Science that was convened by the National Research Council (on which I served) issued a report that took an extremely cautious approach to the use of DNA evidence in criminal trials. At the time, we were concerned that we did not yet know enough about the independence of DNA loci used in forensic testing. In particular, we worried that not enough was yet known about the distribution of variations in the genome across different populations of the human family to permit us to assume that they were inherited independently. We worried that we might be relying too heavily on the so-called "product rule."

Simply put, the product rule supposes that the result from each DNA locus that is tested is truly independent of the results at every other. If that is the case, then one can determine the odds of a random match at each locus and multiply them together. Applying the product rule, the possibility that two DNA samples match merely by luck quickly gets very small, so the argument for the identity of the samples being compared becomes extremely strong. To illustrate, imagine that your DNA has been tested at five loci and that at these five points it exactly matches the DNA profile of some unknown sample. Must the unknown sample belong to you? It depends. If studies of a randomly selected reference population show that the particular type of DNA (in Snowball's case the length of the STRs) you have at each locus is only found in 1 out of every 100 persons, then the odds of the unknown sample deriving from someone else are very small. It is on the order of 1 in 10,000,000,000 ($1/100 \times 1/100 \times 1/100 \times 1/100 \times 1/100$), a number that is less than the total number of people on the planet! Of course if the set of independent STRs in question includes a set of flavors each of which is found randomly in one out of five persons, the odds of a chance match are much higher—about one in 3125 ($1/5 \times 1/5 \times 1/5 \times 1/5 \times 1/5$). This would be a rare event, but, unlike the first calculation, it is imaginable.

How can one prove that the DNA loci selected for the test system are independent? It helps to use markers located on different chromosomes which are by definition not linked, that is, are transmitted independently of each other during the formation of egg and sperm. The real task, however, is to amass a reference population in which one can study the distribution of the various loci and assess whether they are independent of each other. If they are independent, one can compile frequency data and use the product rule as above. The odds of your particular version of a DNA sequence matching that of a sequence taken from a randomly selected sample depends on what reference population was created. A key issue is whether the reference population is an appropriate one to use in determining the odds of a random match. At the least it is necessary to create several populations that account for (usually small) differences in the frequency of DNA loci among major racial groups. Put another way, if you are a member of a subset of the human family that has for centuries tended to marry within a particular subpopulation (say Finns), the standard reference population may not be appropriate to use in calculating the odds that your DNA and a crime scene sample match by chance.

The debate over the degree to which DNA varied at specific points among different ethnic groups waxed hot for several years and generated numerous papers by population geneticists. In the last few years we have learned a lot more about the remarkable similarity of human subgroups when they are compared at most DNA loci and, along the way, most of the concerns about reference populations have been laid to rest. Today, those who want to attack the use of DNA identification evidence in court must focus on whether the laboratory that did the analysis did it correctly (as in the Simpson case). Few, if any, courts will deny admission to DNA evidence based on concerns about population substructure.

When he agreed to attempt to perform DNA identity testing on blood from Snowball and DNA extracted from crime scene cat hairs, Dr. O'Brien realized right away that he faced a feline version of the population substructure problem. Prince Edward Island is a relatively unpopulated, geographically isolated, community. O'Brien knew nothing about the history of the island's cat population. It was certainly possible that most of the cats descended from a small number of founding ancestors. If that was the case, then two randomly selected island cats could well match at several loci. If the island cats were highly inbred, finding a match between Snowball's DNA and that of the crime scene hairs would be of little forensic value because there would be a fair chance that Snowball's DNA would match with that of most other local cat DNA.

If he found that the two forensic samples matched, Dr. O'Brien's task was only beginning. He would still have to satisfy himself so he could eventually testify in court of his certainty that the chances of a match were so low as to allow a jury to conclude that the white hairs on the jacket came from Snowball. O'Brien was able to extract enough DNA from one cat hair to do the identity test. The two samples matched at 10 STRs, double the number normally used in human DNA testing. He next took on the problem of characterizing the genetic diversity of the island's cat population.

The only way to do this was to look at a lot of cat DNA. O'Brien collected blood samples from 19 cats on Prince Edward Island that were thought to be unrelated and from nine cats in different parts of the United States. By studying STRs in these two groups he was able to show that the cats on the island were not highly inbred and that their genetic structure was about as diverse as that of cats in the United States. Once he was convinced that there was sufficient diversity to justify using the product rule, O'Brien used seven newly discovered STRs to calculate the odds that Snowball's DNA matched the DNA from the white hairs found on the bloodstained jacket just by chance. He concluded that the chance was less than 1 in 40,000,000!

O'Brien's forensic report was the crucial piece of evidence in the prosecution's case. There were no witnesses to the crime, no murder weapon was found, and Beamish steadfastly proclaimed his innocence. The DNA evidence placing Shirley Duguay's blood and Snowball's hairs on the same coat was all that convincingly linked Beamish to the crime. The forensic evidence was evaluated by the Supreme Court of Prince Edward Island, which permitted its use. On July 19, 1996, Douglas Beamish was convicted of second-degree murder and sentenced to an 18-year prison term with no possibility of parole.

PCR

The successful prosecution of Edward Beamish depended on the willingness of the Supreme Court of Prince Edward Island to admit evidence of identity that was based on the use in the laboratory of the polymerase chain reaction. PCR was conceived by Kary Mullis, a young, unconventional, molecular biologist, while he was working for Cetus, a biotechnology company in California during the early 1980s. In essence, Mullis realized that there was an astoundingly easy way to use well-characterized and commercially available enzymes to amplify any sample of DNA, however small, until one had as much of it as one needed. This discovery, which garnered Mullis a Nobel Prize in 1992, has revolutionized molecular biology. It has ushered in a new era in cancer diagnostics and in paternity testing, greatly accelerated our efforts to find and clone human genes, and added an important new method to study speciation and biological diversity, to name just a few of its impacts.

PCR involves a sequential series of reactions, repeated over and over. First, the DNA of interest is heated, which causes its two strands to fall apart. Next, very short stretches of single-stranded DNA of known composition are chemically connected to each end of the target DNA. If you could see it, the DNA would now look double-stranded at one end, single-stranded for most of its length, and double-stranded at the other end. The last step is to add the four basic DNA building blocks and an enzyme called DNA polymerase to the mixture. The polymerase recognizes the end of the double strand and chemically adds a new complementary strand to it opposite the middle single strand, continuing until it reaches the other double-stranded region. At the end of the first cycle (which takes minutes), the amount of the original DNA has doubled. As one repeats the series of steps, one keeps doubling the copies of the target DNA. A series of 20 runs amplifies the original amount 1,000,000 fold.

From a forensic perspective, PCR greatly extends the detective's ability to find usable evidentiary samples at a crime scene. It is almost magical in its ability to amplify DNA, a fact that was understood very early in its development. As far back as 1988 in one of the earliest publications pointing out its possible uses in forensics, a scientist demonstrated that he had been able to use PCR to extract and characterize DNA from a single hair root, thus anticipating the work with Snowball. In 1989, scientists showed that they could retrieve small bits of mitochondrial DNA from a 5500-year-old-skeleton and amplify it with PCR.

In 1991, Alec Jeffreys, the British scientist who in 1985 had launched DNA forensics using a different kind of technology, became the first to use PCR analysis of bone DNA to positively identify a murder victim. The badly decomposed remains of a 15-year-old girl had been found wrapped in a carpet in a shallow grave. From study of the skull and dental records, forensic anthropologists had tentatively identified the remains as those of a teenager missing since 1981. Jeffreys extracted the DNA from a small sample taken from the interior of the thigh bone. About 90% of it was actually DNA from soil bacteria, and the human DNA was badly degraded into very short stretches. He used short CA repeats at six different loci and was able to type the DNA. He was then able to compare it to DNA taken from the victim's presumptive parents and confirm the identity, despite the weakness of then-available reference population data. A year later, Dr. Mary-Claire King, a Berkeley scientist who was soon to gain fame for proving the existence of a gene in which mutations predisposed to hereditary breast cancer, became the first to use PCR to establish the identity of a murder victim by analyzing the DNA in teeth.

These first reports created a major stir in the forensics and biotechnology communities. In 1993, Cetus, the company at which Mullis and his colleagues had developed PCR, began marketing a PCR-based kit for studying forensic samples. Within a year many prosecutors were having their first go at using PCR-generated evidence in court.

Especially with regard to criminal law, the courts are conservative in evaluating proposals to admit evidence based on new scientific methods. The courtroom floor is littered with technologies, such as early "lie detector" tests and the paraffin analysis of gun shots, that did not stand up to scrutiny. The justices on the Supreme Court of Prince Edward Island were willing to admit PCR-generated evidence only because by 1996 a sufficient number of other judges around the world had already scrutinized similar evidence, satisfied themselves that there was a firm scientific basis to admit it, and published their opinions.

A murder trial in Massachusetts at which I testified illustrates the power of forensic PCR. In the summer of 1994, a man was bludgeoned to death in a woodshed in the town of Athol, a rundown city located midway between Boston and the Berkshires. The bloodstained murder weapon, a 4-foot-long 2 by 4, lay nearby. Detectives sealed off and painstakingly studied the crime scene. Among the many items they picked up for analysis were several cigarette butts. Forensic analysis found that all the blood at the crime scene derived from the victim. The laboratory next decided to use PCR technology to attempt to characterize DNA from the dried cigarette butts. After using sophisticated methods to extract DNA from the few cells that might be caught in the fibers of filter paper, the laboratory technicians then used a relatively new system of forensic probes developed by Cetus to characterize the DNA. They succeeded in constructing a DNA fingerprint. But the evidence was of value only if it could be linked to a suspect.

The police had a single lead based on a sketchy eyewitness report. On the night of the murder, someone saw a man, whom he could describe moderately well, casually pushing the victim in his wheelchair along a street near the crime scene. The report suggested that the killer knew his victim, which is often the case, and gave the police the rationale they needed to focus their investigation on the whereabouts that night of just a few persons. They eventually arrested a young man, charged him with the crime, and obtained a court order to obtain his blood for DNA analysis. Forensic analysis matched his DNA with the results of the cigarette butt studies. However, the prosecution still had a problem. The suspect, who acknowledged knowing the victim, claimed that he had been at the crime scene many times in the past, and had dropped cigarettes there on many occasions. He did not have a strong alibi for his whereabouts at the time of the murder.

I became involved in the case at the eleventh hour. Late one afternoon the court-appointed defense attorney, who had been searching for weeks for an expert to challenge the use of DNA evidence generated by PCR amplification, got to the end of his list and called me. I was extremely reluctant. Because I am a clinical geneticist and had served on the national committee that issued the first major report on DNA forensics (one that barely mentioned PCR, which was then still in its infancy), I knew that I would qualify as an expert. But I had little laboratory experience with DNA testing and I had always avoided involvement in criminal trials. However, the defense attorney, an engaging former high school teacher named Harry Miles, whom I came to respect tremendously for his courtroom skills, won me over with an argument that I could not refute. He had no more names, the trial was to start the following week, and without an expert to at least raise some doubts about the PCR evidence, he could not provide his client with a full defense. He passionately believed that our criminal justice system could not function fairly if he could not produce someone to raise questions about PCR.

Thus, I found myself driving along route 2 through the beautiful little towns of north central Massachusetts, to a picture-postcard New England courthouse, rehearsing my arguments against a technology I essentially believed in. Miles wanted me available to rebut at least some of the prosecution's expert testimony. There was no small amount of irony here. The prosecution's expert was a young molecular biologist who had been working at the state forensic lab in Connecticut for about a year. His boss, Dr. Henry Lee, had sent him to testify in *support* of DNA evidence. Dr. Lee, a celebrated forensic expert who had worked on the national DNA forensics committee with me, was unavailable because he was in California at the O.J. Simpson trial testifying *against* the use of DNA evidence! At that time, the fall of 1995, the interminable Simpson case had been considering DNA evidence for weeks.

Miles knew that he was asking someone who generally believed that DNA evidence was accurate and trustworthy to testify against it. I was willing to address only certain issues. In 1995 there were still some unresolved technical problems, mostly concerning the adequacy of the reference populations used to calculate the odds that DNA from a randomly selected (innocent) person would match the crime scene sample. I was comfortable discussing those. Furthermore, no forensic lab yet had much courtroom experience with PCR. The prosecution was using it for the first time, the defense was fighting it for the first time, and a Massachusetts judge was evaluating it for the first time.

As it turned out, the evidence in question was so novel that even my small attack on its validity was enough to help make the assistant district attorney tack in the direction that attorney Miles wanted. Before trial she had steadfastly refused to plea-bargain away from a murder indictment, and the defendant had resolutely proclaimed his innocence. A few days after the expert testimony, however, she accepted the defendant's offer to plead guilty to the lesser charge of manslaughter.

Since the early days of courtroom PCR, evidence generated by it has been used in several thousand trials in the United States. Only a handful of courts have refused to admit PCR and those almost always due to concerns about how a particular forensic lab did or recorded its analysis. In 1997 the Supreme Judicial Court of Massachusetts, a court that took a very cautious approach to forensic DNA evidence in the late 1980s and early 1990s, ruled that PCR-based evidence was admissible. Since then, courts in many other states have followed suit.

As scientists have become more adept at using PCR to amplify minuscule amounts of DNA, they have greatly extended its original application. In 1997, Roland Van Oorschot, a DNA forensic geneticist in Australia, really pushed the envelope. A knife that police thought had been used in an attempted murder was given to his lab for analysis. Although it did not have any blood on it, Van Oorschot reasoned that someone must have held it, so he swabbed the knife down and ran the standard PCR protocol. He came up with sufficient DNA to construct a DNA profile. This led him to hypothesize that people leave sufficient DNA on virtually whatever they touch to permit a lab to develop an identification profile.

To test the hypothesis, Van Oorschot and his colleagues tried to obtain DNA from common objects. Using cotton cloth moistened with sterile water and handled with sterile forceps, they swabbed and studied briefcase handles, pens, a car key, a mug, a glass, a telephone handset, gloves, and the insides of four condoms which had been worn but into which there had been no ejaculation (a study that is relevant to the investigation of a significant fraction of rapes). They also swabbed hands. Virtually all the objects yielded enough DNA to identify the user. The scientists even swabbed the hands of men whom had recently shaken the hands of other persons. They showed that merely by shaking hands there is a transfer of DNA. That is, under controlled circumstances they were able to identify who had most recently shaken hands with the person whose hand was swabbed and studied. They also showed that it is possible to obtain DNA from the fingerprints left on glass.

The Australian detectives and colleagues in many nations are now taking a new approach to crime scene investigation. They are assuming that every criminal leaves DNA at the scene. When evidence is gathered properly, the yield has been astounding. For example, forensic scientists have identified DNA obtained from inside a glove left at a burglary, from the rim of a glass, and even from a fingerprint on a window sill. Detectives in England have used DNA to revolutionize the approach to solving auto theft. They now routinely swab the steering wheels of recovered stolen vehicles. DNA analysis typically yields two or more profiles. Of course, one almost always belongs to the owner. The second profile is often that of the thief who drove the car for a while and then abandoned it. Oddly enough, no one is quite sure where the DNA comes from. The best guess is that it comes from keratinized epithelial cells that break open on the surface of the skin, but these cells have very tough coats, so scientists are dubious. Some think that naked DNA may actually be extruded through sweat glands.

The fact that it may be virtually impossible to be involved with a crime without somehow leaving DNA on some object connected with it triggers immensely challenging questions about the right of the state to use DNA analysis to solve crimes and the right of the citizen to maintain his genetic privacy. Our society may decide that the potential gains in crime prevention constitute an overwhelming argument in favor of mandatory universal DNA profiling at birth. If people shed DNA wherever they go and if there is a routine way to identify crime scene DNA by running the profile against a population-wide directory, it will almost always be possible to identify a person the first time he commits a felony, an extremely important development, given the number of serial criminals. DNA analysis of crime scene evidence will often find biological samples of innocent people, but, in most cases, it should be possible quickly to eliminate them as suspects. Merely showing that one's DNA was found at a crime scene or in connection with it would rarely be enough evidence to convict, but it could often provide strong circumstantial support to the prosecution's case.

Universal DNA banking seems inevitable. Its value in assisting the resolution of virtually any kind of serious crime will, I think, prove irresistible. The speed with which European nations and the various states in the United States have initiated DNA databanks on convicted felons is a harbinger of the future. One can anticipate a sharp debate over whether mandatory DNA sampling for identification purposes violates the constitutional right to privacy, but, in the end, all our DNA profiles will be in a computerized database. In the meantime, there appears to be little reason to wear gloves when committing a burglary; one's DNA is being shed from other bodily surfaces even as one moves.

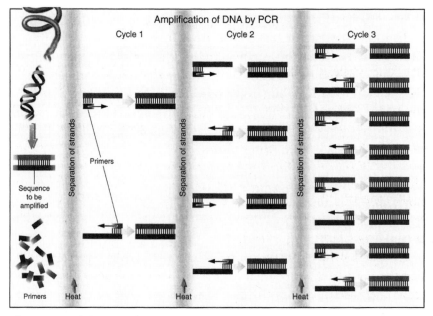

Illustration of the amplification of DNA by the polymerase chain reaction (PCR).
(Reprinted, with permission, from Rosenthal 1994 © Massachusetts Medical Society. All rights reserved.)

6

Cold Hits
The Rise of DNA Felon Databanks

A Murder in Minneapolis

In the fall of 1991, Jean Broderick, a 23-year-old woman just out of Macalester College, decided to share a duplex apartment with Erica Norris, one of her college friends, and two other roommates in Minneapolis. The Lowry Hill neighborhood was a bit shabby, enough to keep the rent low, but not so seedy as to make the three young women and one man worry much about safety. One Saturday night that November, Jean and Erica went out for a few drinks. When they came home they talked for a bit with Erik Helgen, the sole man in the group. That night he had seen the movie *Cape Fear,* the grisly tale of a paroled criminal who terrorizes a family. They talked briefly about how disturbing some of the scenes were, but the three soon went to their respective bedrooms.

Sunday morning started slowly for Erica. She got up about 11:30, tidied the apartment a bit, and then fell back asleep until the phone woke her at 1:30. The call was for Erik, but she knew he was not in because he had said the night before that he was getting up early to attend a day-long retreat for his job. Realizing how late it was and seeing Jean's door open, Erica went in to see her. Jean was facedown in bed with her hands tied behind her back. Her face was an ugly purplish color, and she was stone cold. The autopsy showed that she had been raped and strangled.

For a few horrible hours Erik was the suspect, partly because of the bizarre coincidence that he had seen *Cape Fear* the night before and that Erica had told the police that he had seemed powerfully affected by it. In that movie the criminal picks up a woman in a bar, brings her home, ties her hands behind her back, and brutally beats her. But Erik's alibi quickly checked out; he had been with several people during the morning hours that the coroner determined was the time of the attack. Because the other people in the apartment had not heard any unusual sounds suggesting a forced entry, detectives were convinced that the killer was an experienced criminal. When the police realized that on her last night alive Jean had walked past a halfway house for paroled sex offenders, they focused their investigation there.

In 1989 Minnesota became one of the first states to enact a law creating a DNA felon databank. In Minnesota, before he is paroled, every convicted sex offender must provide a blood sample for DNA analysis. During the mid-1990s technicians at the state Bureau of Criminal Apprehension (BCA) abandoned an older technique called RFLP analysis in favor of a newer DNA identification technology that relied on protocols perfected by FBI scientists. The standard practice was to extract the DNA and use five different DNA probes to type DNA sequences, each of which varies greatly in size among individuals. The result was a genetic fingerprint of each person. The high degree of variability in these DNA sequences is not incompatible with the rigorous evolutionary pruning of genes. The DNA used for identification lies between the stretches of DNA that code for structural genes—the ones that are blueprints for proteins. Because these regions do not code for proteins, the variations are harmless.

The detectives asked the BCA lab to compare the DNA from the semen sample that had been scraped from Jean Broderick's inner thigh with DNA patterns on file for the paroled offenders who were living at the halfway house. There was no match. The DNA evidence that they had been forced to provide at parole effectively eliminated them as suspects. There were no others. Jean had been a quiet person. She had not been dating, and none in her small circle of friends could offer a clue as to who might want to harm her.

When the police asked Joel Kohout, a seasoned detective who works at BCA, to step in, she quickly decided that the crime looked like the work of a recidivist, someone who had raped before. After agreeing that the case was going nowhere, Kohout and her lab colleagues decided to do something that had never been done before. They would compare the DNA profile of the semen sample to the DNA profiles of every convicted rapist in their new DNA felon databank. At the time, the BCA had not yet set up a computerized database of its DNA profiles, so the lab technicians had to visually compare the physical profile of the crime scene sample (essentially, a series of dark bands on an X-ray film) to those in the database. In the first pass they were able to exclude all but six of the men who were registered in the database. Put another way, they found only six convicted sex offenders who matched the sample at the first DNA position they had picked. After running a comparison of a second DNA locus from the crime scene sample against the six remaining open files, the technicians could exclude five more.

The chances that a man randomly selected from the population of Minnesota will have a DNA profile that matches the first two DNA loci in the crime scene sample is less than 1 in 1000. But there are more than two million adult men in Minnesota, so if one tested the DNA of every one of them with two probes, about 2000 would not be excluded as suspects. This is why the FBI insisted on a multi-marker test system (in 2000 a system of 13 STRs was in use). As the number of markers that fail to exclude a suspect rises, the possibility that the match is due to chance alone falls quickly to unimaginably small numbers. This, of course, assumes that the testing laboratory has not made some gross blunder such as mislabeling a sample or testing the same sample twice, or committed fraud—the kinds of concerns that were raised by the defense at the trial of O.J. Simpson.

Jim Liberty, one of the forensic technicians, knew that he had a very strong lead, but he needed to be as certain as possible if the lab work was going to provide the sole basis for an arrest warrant. Obeying protocol, he took care not to learn anything about the sole remaining suspect. Other forensic tests had shown that the semen was from a man with a type of enzyme called PGM 2-. Liberty asked another technician to retrieve the original blood sample of the convicted felon and test it for PGM 2-, a type found in only about 1 in 200 persons. It also was PGM 2-. The chances of a randomly selected man having two DNA loci and the same PGM profile were very small, perhaps 1 in 200,000.

The Minneapolis police had made a "cold hit," a first in U.S. forensic history. By comparing the DNA profile of a crime scene sample to those in an existing database, authorities had, despite the lack of any other evidence, found their man. His name is Martin Estrada Perez, a 37-year-old career criminal. He was easy to find. He was in the Hennepin County jail where he had been since being arrested for burglary just 11 days after Jean Broderick's murder. A test of his blood showed that his DNA matched that of the DNA in the semen found on the dead woman. As the prosecutors prepared for trial, they found other evidence, including eyewitnesses who picked Perez from a lineup as someone they had seen in the neighborhood. On April 23, 1993, a jury, largely on the basis of the cold hit, convicted him of first-degree murder. He was sentenced to life in prison without the possibility of parole.

The Rise of DNA Evidence

Ironically, the scientific basis for matching DNA samples from crime scenes began with the most peaceful of interests. In the early 1980s, Alec Jeffreys, a young molecular and evolutionary biologist working at the University of Leicester in England, was trying to use DNA markers to study the relative reproductive success of animals and how that related to variation in the structure of local populations. Among his first subjects were sparrows. By chance he discovered that there were many highly variable regions in the sparrow DNA that he could use for this purpose. About this time, police in England were trying to solve two brutal rape murders in two villages near Leicester. Hearing of Jeffreys' DNA identification work in birds, they wondered whether he could do it in humans. They were especially interested to find whether the DNA from the semen samples taken from the victims showed that one man had committed both crimes. Jeffreys agreed to try.

Early on, Richard Buckland, a teenage boy with a history of sexual deviance, had confessed to the crimes, but the police doubted his word. Jeffrey's studies showed conclusively that the DNA pattern of this emotionally disturbed adolescent did *not* match that of the semen left by the killer. Thus, the very first use of DNA testing in a criminal case exonerated a suspect who had confessed to a crime he did not commit!

The British police asked all the men who lived in or about the three local villages closest to the crime scenes to voluntarily provide a blood sample for DNA studies. More than 98% (3653 men between 13 and 34) agreed. Not surprisingly, since what killer would give blood, DNA analysis detected no sample that matched the crime scene profile. It did, however, eliminate a huge number of suspects. It also found something extremely interesting: Two of the DNA samples from the volunteers matched perfectly. Since there were no identical twins, this meant that a sample from one person had been analyzed twice. Someone had provided blood under his own name and then again on behalf of someone else. When the police approached the individual who seemed to be the source of two samples, he readily admitted that he had given blood twice, once for himself and once for a friend who had persuaded him to do so because he claimed to be terrified of needles and would be embarrassed if he fainted. The police now had a real suspect. The other man, who had prior brushes with the law for child molestation, eventually confessed.

News of the case electrified the police forces in Europe and the United States, and for good reason. Each year in the United States there are about 20,000 murders and 90,000 reported rapes (almost certainly another 90,000 go unreported). Only a fraction of these cases end in the conviction of the perpetrator; in the case of rape, much less than one-half. Yet, we know that many crimes are committed by recidivists, those who have been convicted before for the same or similar acts. For example, of convicted felons released in 1993, 62% were rearrested within three years of parole and 41% were reimprisoned.

With the exception of murder, criminals tend to commit the same crimes repeatedly during the course of their lives. This is particularly so in the case of convicted sex offenders. The high rate of recidivism provided the fundamental argument in favor of creating DNA felon databanks that swept through law enforcement circles and legislatures in the United States and Europe starting in the late 1980s. During the last decade, all 50 states have enacted laws to create DNA felon databanks. Most laws originally targeted sex offenders, criminals who comprise about 10% of all convicted felons. A few states, such as Virginia, decided early on to collect blood at parole from all persons who had been convicted of a felony. By 1999, it had become clear that most states would revise their laws to follow Virginia's lead. As a result of its decision to enact a databanking statute with broad reach, Virginia has already collected several hundred thousand samples, most of which are awaiting DNA analysis. Why? Paul Ferrara, who directs the excellent, if overburdened, state lab, only has enough resources to analyze samples from convicted murderers, rapists, and other sex offenders. Samples from less dangerous persons must wait.

In the late 1980s when courts were first considering DNA evidence, judges, who knew nothing about genetics, had three major concerns: (1) Was the underlying theoretical basis upon which the effort to match samples rested sound? (2) Was the testing technology accurate? (3) Did the particular laboratory meet the standard of care in testing a particular set of samples? Today, the first two questions have been settled in favor of DNA technology. In courts now the focus is mainly on the quality of the analysis done by a particular lab.

One of the major public policy issues in the criminal justice system today is how to gear state forensic labs up to do DNA typing on the ever-swelling number of tissue samples taken from convicted felons, but not processed. An unanalyzed sample is worthless. It provides no information to the databank. If a paroled individual rapes again, DNA analysis will not lead the police to him. In the U.S., the backlog of unprocessed and uncollected samples is already over 1,000,000 and growing rapidly. In addition, there are an estimated 180,000 "rape kits" (most of which are likely to contain semen from the rapist) stored in evidence lockers around the nation that have never been subjected to DNA analysis. Since rapists typically rape repeatedly, failure to analyze these samples is a national disgrace, and an offense against all women. In late 1999, Congress appropriated $15,000,000 to help law enforcement agencies analyze the huge backlog of unprocessed samples. This is welcome news, but we need at least another $30,000,000 to complete the job.

Despite current funding problems, DNA forensics will forever change the operation of criminal justice. England is leading the way. In August 1994, British Home Secretary, Michael Howard, announced that Britain would build the world's most comprehensive DNA forensic database. Authorized by new legislation, the Home Office has created a computerized national register of the DNA profiles of all those convicted of a felony, a master list to which every future crime scene sample will be compared. In addition, the law permits the police to construct a DNA identity profile of everyone *arrested* for crimes ranging from shoplifting to murder. Without obtaining consent, the police may take a blood, saliva, or hair sample for that purpose from any person who has been arrested. The DNA profiles of individuals who are exonerated are quickly expunged from the database. By extending the reach of DNA sampling to persons arrested for even relatively minor crimes, the British government has taken DNA banking to a new level and drawn fire from civil liberties groups.

In the United States there has, thus far, been relatively little debate about DNA felon databanks, probably because in most states until recently the programs have only collected samples from persons convicted of truly heinous crimes. Convicted felons have challenged the laws in at least nine states, including Virginia, arguing that they are an illegal invasion of privacy and violate the constitutional rule which forbids punishing a person pursuant to a law enacted after he was convicted of a particular crime. The courts have consistently upheld the DNA felon databanking laws as a reasonable exercise of the police power that does not violate the search and seizure provision of the Fourth Amendment. No court has yet refused to admit into evidence testimony that a DNA analysis of a sample collected from a convicted felon matched that of DNA from tissue at a crime scene.

Despite the fact that the key constitutional issues may already be settled, several contentious issues remain that are likely to be resolved only in court. For what criminal behavior is it appropriate to require admission to the DNA felon databank? Should we follow England in taking a broad approach, or does our differing legal approach to privacy require a narrower reach? After preparing a DNA analysis, should we retain the tissue sample or destroy it? The major argument favoring retention is that testing technologies evolve and that samples may have to be retested to conform to some new testing algorithm. This seems much less an issue than it was earlier on. It is highly likely that the current, federally organized system that uses 13 STRs is sufficient for at least a decade, even with the advent of more powerful tools. The major argument against retaining the sample is rooted in suspicion that the state will not protect the privacy of the samples. Theoretically, one could ask a vast number of questions about the sample, the answers to which would reveal many other facts about the convicted felon. For example, one day we may be able to ask whether he is genetically predisposed to alcohol abuse. Should the state be allowed to do that? The laws in most states currently forbid such inquiries unless the research is conducted anonymously. Another, related, issue concerns the length of time samples should be retained. Some favor a long term (e.g., 50 years), whereas others prefer a shorter period. A fair compromise might be a 10-year holding period; long enough to accommodate a new technology, but short enough to reassure those who are worried about a potential loss of privacy.

A key measure of a criminal justice system is how it treats the accused. It is reassuring that DNA felon databanking and evidence testing are a boon to the defense. When DNA testing exonerates a suspect, it does so absolutely. One early FBI estimate was that DNA testing redirects the investigation away from a leading suspect about 30% of the time. Even more important is the use of DNA testing to reopen old cases and free wrongfully convicted people. An FBI official who did not want to be quoted told me that he guessed that 5% of men in jail for rape were innocent of the crime for which they were convicted. In the minds of some criminal defense attorneys this is a conservative guess.

Around the country there are scores of efforts under way to reopen cases for DNA testing. Barry Sheck, one of the two DNA lawyers on the O.J. defense team, runs the "Innocence Project" at Benjamin N. Cardozo School of Law in New York. He and his colleagues are looking into hundreds of cases in which convicted rapists seek DNA testing to challenge their convictions. In 1995 he and Peter Neufeld, the other O.J. lawyer who attacked the DNA evidence, successfully convinced Westchester County to release Terry Leon Chalmers, who had served eight years for rape on a conviction based largely on the victim picking him out of a police lineup. DNA testing, which had not been done in the first trial, excluded him. In the words of Jeanine Pirro, the District Attorney, DNA is "like the finger of God pointing down saying 'You did it'" ... or, in this case, that you did not do it.

The number of incarcerated felons who have regained their freedom because a DNA analysis obtained by the Innocence Project showed that the convicted person was not the source of the semen or blood stain found on the victim is growing steadily. In 1999 Professor Sheck knew of 53 wrongfully convicted persons who had been freed, one when he was only days from execution. The number might be much higher if it were not for laws that sharply limit a duly convicted person's right to reopen the case.

THE SHEPPARD MURDER: USING DNA TO REVERSE JUDGMENT

The most famous use of forensic DNA testing to help a wrongfully convicted man arose 26 years *after* his death. Probably ranking just behind the 1935 trial of Bruno Hauptmann for the kidnap and murder of Charles and Anne Lindbergh's son and the 1995 trial of O.J. Simpson for the murder of his former wife and her friend, Ron Goldman, the 1955 trial of Cleveland surgeon Sam Sheppard for his wife's brutal murder was one of the crime stories of the century. Sheppard claimed that on the night of July 4, 1954 he was asleep elsewhere in the house and awoke to his wife's screams.

When he rushed upstairs he was knocked out briefly by a tall "bushy haired intruder" whom, after he regained consciousness, he chased toward the waters of Lake Erie near his lakeside home. There he claimed the killer again overpowered him. But Dr. Sheppard's credibility was greatly harmed when a woman challenged his testimony that he was happily married. Susan Hayes, a hospital lab technician, testified that in the weeks before the trial she had sex with him on several occasions in his automobile.

Sheppard was convicted and served 10 years in prison for his wife's murder until he successfully overturned the conviction, largely on the basis of a review of the trial transcript, after which the appellate court concluded that the trial judge had let a carnival-like atmosphere prevail, which could have easily prejudiced the jury. A key fact that the prosecution had never adequately dealt with at the original trial was the long trail of blood that was found in the house. The prosecution said it had dripped from the weapon that Sheppard had used to bludgeon his wife and that, since Sheppard had no wounds, it was her blood. But in 1954 there was no good means to test it.

Although Dr. Sheppard died in 1970, his son, who was only seven when his mother was killed, never stopped trying to clear his father's name. At times he has seemed close to victory. In 1996 Mohammad Tahir, an accomplished forensic scientist in Indianapolis who had volunteered to work with Dr. Sheppard's son, determined conclusively that the blood stains throughout the house were not from Mrs. Sheppard. He had been able to test strands of hair that had been taken from the bed in which she had been murdered and compare them to the blood. Since the blood was not Dr. Sheppard's and it did not match the DNA from her hair, a third person must have been present at the crime scene.

In 1959 a man named Richard Eberling, who had washed windows in the Sheppard home, came under suspicion when a diamond ring that had belonged to Mrs. Sheppard was found in his possession. He admitted stealing it, but said the theft had been from Dr. Sheppard's brother. In 1989 Eberling was convicted of murdering a 90-year-old woman after forging her will. He died in 1998 while serving a life prison term.

Although it was not brought out at trial, Mrs. Sheppard was found with her pajamas at her feet, suggesting the possibility of rape. More than 40 years after they were prepared, slides of two vaginal swabs that had been taken from Mrs. Sheppard the night of her death were examined by Dr. Tahir. He found sperm cells. Using PCR technology, he was able to get enough DNA for testing. In his report to Ohio officials, Dr. Tahir said that the testing was "not incompatible with Eberling's DNA profile." Elizabeth Balraj, the Cuyahoga County coroner, announced that she would not challenge Dr. Tahir's findings.

In a poignant coda to the Sheppard murder, the couple's son, who is absolutely convinced of his father's innocence and certain that Eberling murdered her, arranged in 1997 for his father's remains to be buried with those of his wife. At the exhumation, he obtained hair samples for DNA analysis that he claims absolutely excluded Dr. Sheppard as the source of the semen sample found on his murdered wife. There the matter might have rested had not Samuel Sheppard filed a wrongful imprisonment suit against the state of Ohio. Nothing is more embarrassing to prosecutors than an allegation that they sent the wrong person to prison. In the fall of 1999, as part of his preparation to defend the wrongful imprisonment lawsuit, Cuyahoga County Prosecutor William Mason won permission from the court to exhume Mrs. Sheppard in order to obtain tissue samples for extensive DNA analysis. Among other things, he hopes to show that Dr. Tahir's forensic work on hair purporting to be from Mrs. Sheppard is in error due to sample contamination. If this work reaches the same results as did Dr. Tahir's analysis, it will surely help the son's claim of his father's innocence.

OTHER USES OF DNA DATABANKS

In addition to championing the development of state-based DNA databanks to compare crime scene samples to archived samples, the FBI asserts that DNA databanks provide at least three other important benefits. First, in studying a crime scene, detectives can use DNA analysis to determine whether blood or semen stains derive from one or more individuals. Second, in communities that have been terrorized by a wave of unsolved crimes such as rape, by comparing samples from different crimes, forensic scientists can determine whether the violence has been committed by several different men or is the work of a serial criminal. The resolution of this issue can be extremely helpful in how other evidence is weighed. Finally,

because the molecule is so stable, DNA analysis can be used to help identify badly decomposed human remains.

In the United States this kind of forensic work was first used on a large
scale by the Department of Defense during Operation Desert Storm to reassemble body parts of U.S. personnel who were killed when a missile hit
their barracks. It has proved extremely helpful in civilian air disasters such
as the crash of TWA flight 800. In September, 1997, the DNA lab at the
Armed Forces Institute of Pathology finished the task of identifying every
person who died in that tragedy. The last 15 tests used DNA obtained from
tiny bone fragments dredged from the ocean floor. Over the last decade,
scientists have used the fact that close relatives have a high probability of
sharing DNA sequences to return the bones of the "disappeared," victims
of political assassinations in Chile, to their families. Since many of the victims were young parents whose small children were kidnapped and subjected to forced adoption, DNA analysis has also been used to reunite children with their true grandparents.

In 1990 the Department of Defense decided to create a set of DNA reference samples on all members of the United States military. In the solemn
words of Lieutenant Colonel Victor Weedn, the pathologist and DNA expert who until 1997 directed the DNA Identification Lab at the Armed
Forces Institute of Pathology in Washington, "We never want to place another set of human remains in the Tomb of the Unknown Soldier at Ar-
lington National Cemetery." The U.S. Department of Defense now operates the world's largest DNA databank. In 1999 it contained tissue samples
of more than 3,000,000 individuals. Each new recruit provides two blood
samples and a saliva sample that are dried and stored separately in two locations. DNA is only actually analyzed if a problem in identifying human
remains emerges. Currently, the plan is to store the samples for 50 years,
but, upon discharge, a soldier may request that the sample be destroyed.

Advances in DNA forensics, such as the ability to study old evidence as
has been done in Sheppard case, have caused some novel legal problems.
In the vast majority of states it is quite difficult to reopen a case after final
judgment. However, in cases where it is possible to perform DNA analysis
on evidence that was *not* analyzed for trial and in which the results could
exonerate a wrongfully convicted person, the individual should have the
right to obtain such tests and win his freedom if the results unequivocally

exclude him as the source of the DNA. We need to change the rules for post-conviction appeals to accommodate important DNA evidence, and justice demands that we do it at once.

DNA forensics also will force us to reexamine the use of statutes of limitations, the laws that forbid an individual from being tried for a crime after a number of years have elapsed. In October, 1999, with the tolling of the six-year statute of limitations on a 1993 rape only days away, Milwaukee Assistant District Attorney Norman Gahn filed an arrest warrant against one "John Doe, unknown male with matching deoxyribonucleic acid profile." Gahn found that the DNA profile in question matched that constructed from semen stains obtained from two other victims in Milwaukee who were raped about the same time. He convinced a judge that a DNA profile is better than an alias or a written physical description, both of which often provide the basis for a warrant. He has placed the DNA profile in the state DNA felon databank, and he is confident that it is only a matter of time before it turns up as a match with DNA from a contemporary crime. This will make a long-cold trail suddenly quite warm.

The advent of DNA databanks such as that of the Department of Defense, which will eventually hold tens of millions of samples, makes one wonder where we are heading. In 1998 Louisiana became the first state to enact a law permitting the taking of a tissue sample for DNA identification at arrest. Early in 1999, New York Mayor Ralph Giuliani and Police Commissioner Howard Safir publicly advocated mandatory DNA testing at arrest (a policy which if implemented would lead to 300,000 tests a year in New York City alone). In November 1999, the International Association of Police Chiefs announced that it will urge Congress to require that DNA samples be taken from every person who is arrested. Since the technology is already in place to create DNA profiles on everyone, should we do so?

For decades we have been collecting blood samples from newborns to screen them for evidence of rare, treatable disorders which, unless detected quickly and properly managed, can lead to mental retardation. It will soon be relatively inexpensive to prepare a DNA profile on each child as well. Should we? Those who favor the creation of a universal DNA identity bank argue that if we maintain a DNA profile on each person, career criminals are much more likely to be apprehended early, sharply reducing the number of crimes they commit before they are caught. Others, especially in the

United States, view such proposals as foreshadowing the rise of an Orwellian state.

In some respects, a system to compile the DNA profile of every person would be much more fair than the current approach. In limiting DNA profiling to convicted felons we are constructing a bank whose membership will reflect current social prejudices. If, for example, blacks are more likely than whites to be convicted of a particular offense, then they will in a sense be overrepresented in the bank. The bank in turn will be of greater value in apprehending black perpetrators of future crimes than white criminals.

DNA felon databanks are here to stay. One can only guess about the future uses to which they will be put. One cause for concern is that the banks currently store whole DNA, a practice that is not necessary to accomplish the core purpose of using the banked DNA as reference samples against which to try to identify a crime sample. By retaining whole DNA on convicted felons, especially certain groups such as sex offenders or persons whose alcohol abuse led them to commit vehicular homicide, we are creating databanks that will be of immense potential interest to behavioral geneticists. If we determine that certain genetic variants are much more common among convicted felons than among the general population, it is possible that some will argue that these are markers of predisposition to commit crimes. What will we do with such information? It is likely that tests showing the presence of such genetic variants could influence decisions about parole. A parole board might well conclude that a felon with the genetic predisposition to a particular behavior is more likely again to commit crimes than a felon who does not carry the marker. Some scientists will also find it irresistibly interesting to try to identify children with such predisposing genes and track their development and behavior.

Although I support the use of DNA databanks to help apprehend criminals and exonerate innocent suspects, I oppose the use of this archived DNA for prospecting in behavioral genetics. Even in the best of circumstances, such studies are likely to yield only mildly impressive correlations. But such findings could mislead many in our society to embrace the foolish notion that crime is a function of biology, thus directing it away from the immense socioeconomic problems that breed lawlessness.

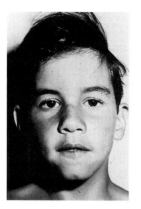

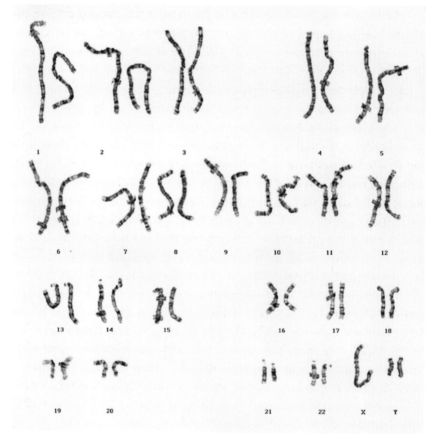

Eight-year-old boy evaluated for behavioral problems and learning disabilities. The prominent glabella and long face are compatible with the presence of an extra Y chromosome (47, XYY syndrome). (Photo reprinted, with permission, from Jones 1988.) (Karyotype courtesy of Genzyme Genetics.)

7

Genes and Violence
Do Mutations Cause Crime?

THE XYY SYNDROME

On Christmas day, 1965, Patricia Jacobs, a promising young cytogeneticist, and her colleagues published an astounding report of their studies of the chromosomes of 197 men who had been committed to the Carstairs Hospital for the criminally insane in Scotland.

The 1960s was the golden age of cytogenetics, a period when, after a dark age, our knowledge of human chromosomes grew immensely. Since 1902, the year that Sutton and Boveri proposed that chromosomes carried the hereditary material (the word gene was not yet in use), cell biologists had labored to understand them. In the case of human tissue, however, their tools were of such limited value that for more than 30 years scientists had accepted an erroneous report by Theodosius Painter that each human cell had 48 chromosomes (23 pairs of autosomes and 1 pair of sex chromosomes). In the early 1920s when Painter did his work, it was not possible to count chromosomes (which under the microscope looked like a plate of spaghetti) in most types of tissue. Knowing that the cell replication cycle of germ cells made them a better object of study, Painter sought permission to dissect the testicles of two inmates of the Texas State Insane Asylum whom authorities had castrated in an effort to control their behavior. In his first paper, he suggested the correct number was either 46 or 48; in his second, he asserted that 48 was the correct count. It was not until 1956 when Tio and Levan, two scientists working in New York, developed a new method to spread out chromosomes so fewer overlay each other that the correct count was discovered to be 46.

In 1958, the French geneticist, Jerome Lejeune, discovered that persons with Down syndrome had 47 chromosomes (the normal complement plus an extra number 21), thus confirming a hypothesis about the cause of Down syndrome that had been proposed nearly 30 years earlier. This extremely important discovery triggered huge interest among other scientists. Over the next few years, other researchers made rapid progress in finding abnormal numbers of chromosomes and associating them with physical or mental abnormalities. But no one was prepared for Dr. Jacobs' report.

Jacobs was acting on a hunch. She knew that another cytogeneticist had recently counted the chromosomes in white blood cells taken from 942 men housed in English institutions for criminal and/or mentally re-tarded men. He had found that 21 of them had an extra X chromosome, and that of these, 7 also had an extra Y chromosome. These findings con-trasted sharply with a large study of 2607 mentally subnormal men with-out significant behavioral problems of whom only 2 had been shown to have the extra Y. Jacobs and her associates wondered "whether an extra Y chromosome predisposes its carriers to unusually aggressive behaviour." If this were true, one would expect to find an unusually large number of men with an extra Y among groups of men with a history of unusually violent behavior.

The Carstairs Hospital housed 203 men, of whom 197 agreed to un-dergo study. Among them, Jacobs and her colleagues found 12 men with an abnormal karyotype (chromosome count), of whom 7 had an extra Y chromosome. At the time, almost nothing was known about the physical or clinical consequences of being born with an extra Y, but the finding that 3.5% of the Carstairs Hospital population had XYY syndrome was with-out question much higher than would be expected in a random sample of the normal male population. For example, Jacobs had studied 1500 appar-ently normal men and found only one with an extra Y. If one assumed that about 1 in 1000 newborn boys had an extra Y, the Carstairs group had more than 30 times the expected number. The other interesting finding that emerged from the Carstairs study was that men with an extra Y chro-mosome were very likely to be much taller than their counterparts. Those men had an average height that was a full 6 inches greater than the other 190.

Jacobs entitled her brief paper: "Aggressive Behaviour, Mental Sub-normality and the XYY male." She closed the report on a provocative note, writing: "At present it is not clear whether the increased frequency of XYY males found in this institution is related to their aggressive behaviour or to their mental deficiency or to a combination of these factors." The implication was clear; in her opinion, the XYY syndrome might include an inborn tendency to aggressive behavior. Thus was born the notion of the "criminal chromosome," an idea that she did not propagate, but which fascinated the world's journalists.

The report generated intense interest and spirited (sometimes bitter) debate in the scientific and forensic community. As clinical geneticists found and studied more men with XYY syndrome, they learned that in addition to being tall, some of them had coarse facial features, scars from severe acne, and low intelligence. In a word, many of them looked much like Hollywood's portrayal of a hulking criminal. On the other hand, most men with XYY syndrome looked normal, were clearly of normal intelligence, and were leading ordinary lives. A number of researchers argued that the reason that persons with an extra Y chromosome were more likely to turn up in prison populations was because if they committed petty crimes they were more likely to get caught, and if they were tried before a jury they were, because of their "criminal look," more likely to be convicted.

In 1968, thinking that it might settle the controversy over whether a second Y chromosome predisposed to violent behavior, several physicians at Harvard initiated a project to examine the chromosomes of thousands of newborns and then follow those with the XYY and other abnormal chromosomal constitutions through childhood and adolescence. As word got out, the project drew a storm of protest, especially from a Boston-based group called Science for the People, whose members argued that the study was scientifically worthless and likely to harm the children. In their view, parents of children who were found to have the extra Y were likely, despite the advice of the researchers, to view their child as having a behavioral abnormality. This would create a self-fulfilling prophecy. If the parents treated the children as though they were abnormal it was likely that they would develop behavioral problems. The scientific debate became rancorous and was soon immersed in confrontational politics. Faced with mounting adverse publicity, the physicians abandoned the study. To this day, there have been far fewer studies of the developmental course of children with XYY syndrome than of similar cytogenetic conditions such as Klinefelter syndrome (in which boys are born with an extra X chromosome) or Turner syndrome (in which girls are born with only one X).

The controversy largely subsided after the publication of a major study of XYY men in the prestigious journal, *Science,* in 1976. A team of scientists concluded that, as had been posited, many of the XYY men in prisons had wound up there because borderline intelligence and low socioeconomic status had put them at higher than usual risk of being caught and convicted. They were, for example, less likely than others to have a personal attorney and less likely to obtain a successful plea bargain.

Despite the resolution of the scientific debate, defense attorneys became intrigued with the possibility that they might be able to defend clients on trial for violent crimes by invoking a "criminal chromosome" defense. This started in 1968 when a court in Australia acquitted a man who had murdered a 77-year-old woman on the grounds that he was insane because of XYY syndrome. A few weeks later, a French court convicted Daniel Hugon of murder, but reduced his sentence to 7 years because he had an extra Y chromosome. In the United States there were at least seven murder trials in which the defense tried to win an acquittal or obtain a lighter sentence because the defendant had an extra Y chromosome. Only two judges even permitted the evidence to go before the jury. In each case, the prosecution won a guilty verdict. In the only American murder trial in which the defense was permitted to fully develop an argument that the XYY syndrome should be grounds for acquittal based on the presumption that the individual could not control his behavior, a New York jury convicted a 6-foot, 8-inch defendant named Sean Farley.

The insanity defense, a relatively new addition to Anglo-American law, emerged during the late 19th century. Trying cases in which the defendant had engaged in strikingly abnormal behavior that seemed devoid of motive, courts on both sides of the Atlantic concluded that some persons might be incapable of judging right from wrong. For them the concept of guilt had no meaning. In a closely related line of cases, the courts also recognized that there could be rare instances in which persons could be placed in situations in which it was conceivable that they could suddenly lose all measure of self-control and succumb to an unpremeditated, "irresistible impulse." Throughout its history, the insanity defense has been used sparingly, and although it is widely known through movies and plays, it is rarely successfully invoked in real life.

The idea that the XYY syndrome might actually predispose to behavioral abnormalities that might lead to an application of the insanity defense has been dormant during the last two decades. However, there has been no shortage of creative tactics by defense counsel, especially those involved in murder trials. In the last 20 years, low blood sugar levels, high blood sugar levels, premenstrual syndrome, and posttraumatic stress syndrome have all been offered—almost always unsuccessfully—as a basis for an insanity plea.

We now know that the vast majority of people with XYY syndrome do not commit violent crimes. However, we also know that a significant fraction of people in prison and an even higher percentage on death row are of borderline intelligence or have mild mental retardation, and that many of them have a history dating to early childhood of severe behavioral problems. It is probable that many of these (mostly) men have gene variants that contribute to their low intelligence and to their behavioral problems. It is possible, although by no means proven, that in a few individuals a gene defect may have constituted a major etiologic factor in criminal acts. One recent murder trial shows how tenacious defense attorneys will be in trying to make that case.

Monoamine Oxidase A Deficiency

In February, 1991, Stephen Mobley, a young man with a long history of brushes with the law, shot and killed the manager of a Domino's Pizza store in Oakwood, Georgia. Because he had already cleaned out the cash register and had met no resistance during the robbery, the motive was unclear. In the ensuing month before Mobley was apprehended, he committed six more armed robberies. When he was finally caught, he readily confessed to the murder and, while waiting for trial, often bragged to fellow inmates about it. He kept a Domino's Pizza box in his cell and, reportedly, threatened guards by telling them that they looked like Domino's delivery boys.

At trial, Mobley's attorneys tried an innovative strategy. Instead of portraying him in the best possible light, they emphasized his troubled childhood, his violent past, and his prior convictions. They filed a motion asking the court to order the State to provide funds so that they could obtain testing that might reveal that Mobley suffered from a genetic defect that caused abnormal levels of a key chemical in his brain. If so, they argued, he might well be driven by violent impulses that were beyond his control. The attorneys based their motion on two arguments. First, they provided biographical information about a dozen of Mobley's relatives over four generations. They portrayed a family, many of whose members could not control their behaviors, drank heavily, and often committed crimes.

Their brief is eerily reminiscent of the stories of the Jukes, the Kallikaks, and other families studied by eugenicists a century ago. These now quaint case reports once struck fear into the American public with their lurid depictions of huge families in which a strong propensity to criminality and prostitution were as surely inherited as eye color and the family chin. With only a few exceptions, the attorneys asserted, the Mobley family tree bore similarly poisoned fruit. Specifically, they argued that Stephen Mobley and some of his relatives might well be burdened with the same genetic disorder as was a family in The Netherlands in which a new genetic disease had been diagnosed.

The discovery in The Netherlands began early in 1978 when a young woman walked into a genetics clinic at a hospital in Nijmegen and told a physician that she was worried by the history of mental retardation in her family. She wanted to know if there was a test to show whether she would be at risk for bearing a son with similar problems. As Dr. Hans Brunner, a clinical geneticist, talked with her, he soon realized that the family history of mental retardation was only the tip of the iceberg. Virtually all the men with mental retardation also sometimes behaved with extreme violence. As the months passed, Dr. Brunner was able to review family records that made it impossible to dismiss his concern. He discovered that 30 years earlier, an unaffected granduncle of the woman who had come to him, convinced that there was a family curse, had compiled detailed accounts of all living relatives that he could locate. The granduncle had identified 9 male relatives with mental retardation, all of whom had periodically had bursts of extraordinarily violent behavior, many involving sexual assaults against their sisters. Since then, the family had grown to include five more men who were both mildly mentally retarded and violent.

Brunner realized that the family might well be burdened with a heretofore undescribed genetic disorder. He and his colleagues, with the grateful cooperation of the family, decided to try to find the abnormal gene. Over the years, they were able to find, examine, and draw blood for DNA analysis from 24 family members, 8 of whom were men with mild to borderline mental retardation. All 8 men, who came from four different nuclear families, had behavioral problems. They seemed to undergo periodic bursts of severe aggression lasting for several days during which they slept little and, when sleeping, experienced severe night terrors. One young man had raped his sister. Several years later, while in prison on a work detail, he had stabbed a guard who had reprimanded him for a minor infraction in the chest with a pitchfork. Another man had attempted to run down the supervisor of his sheltered workshop with an automobile after that man had mildly chastised him for laziness. Other men had exposed themselves in public. A third man had on several occasions attempted to rape his sisters at knife point. Two others had committed arson. In all four families, the young women relatives refused to be alone with the men because of their frequent, inappropriate sexual advances.

The family history overwhelmingly suggested that, if the mental retardation and violence were due to a faulty gene, it almost certainly was on the X chromosome. Brunner and his team used DNA markers known to reside on that chromosome to map the location of the gene in which a defect might lie. The goal was to find a marker (a short stretch of repetitive DNA that because of wide variation in length among individuals can be used as a molecular address in an individual) that would distinguish men with the disorder from those who did not have the disorder. Using 26 different markers, the researchers were soon able to identify a region on the long arm of X that was highly likely to contain a faulty gene. That is, they found a DNA marker that was always present in affected men and absent in unaffected men, a fact that strongly suggested that the short stretch of DNA was tightly linked to and co-inherited with the abnormal allele. They next asked which of the genes in that region might, if defective, predispose to such bizarre behavior.

Among the candidate genes known to reside in the suspect region was a gene that codes for a protein called monoamine oxidase A. This protein is found in neurons in the brain and is responsible for regulating the level of important neurotransmitters called catecholamines. Although no one had ever described a patient with a solitary defect in monoamine oxidase A (MAOA), there had been case reports of persons missing chunks of this region of the X chromosome who were severely retarded. Brunner's group tested the urine of the affected men to look for evidence of abnormal levels of MAOA. The results suggested that these men made either very little of this important brain chemical or none at all. The evidence strongly suggested, but did not prove, that these men had a mutation in the gene coding for MAOA.

Just a few months later, Dr. Xandra Breakfield, a molecular biologist at the Massachusetts General Hospital who had agreed to collaborate with Dr. Brunner, found that the men in this family did have a mutation in the MAOA gene, a tiny defect that prevented production of a functional protein. Their paper stands as the first definitive proof that a mutation in a single gene can drive behaviors that are universally considered to be aberrant.

In their petition to the court to permit genetic testing, Mobley's attorneys relied heavily on the work done by Dr. Breakfield and presented an affidavit showing that she had agreed to perform the relevant tests should the court authorize them. The prosecution countered that despite the violent family history, there was little reason to suspect that Mobley was afflicted with a genetic disorder involving MAOA. In Mobley's family, both women and men exhibited violent behavior; furthermore, none was mentally retarded. Stephen Mobley himself had a normal IQ. Also important was the lack of evidence to connect any genetic condition with a propensity to commit any violent crime. After comparing the Mobley family history with the papers written by Brunner, the trial judge rejected the defense motion, asserting that the "theory of genetic connection . . . is not at a level of scientific acceptance that would justify its admission." On February 20, 1994, a jury found Mobley guilty of murder and sentenced him to death. As part of his appeal to the Supreme Court of Georgia, Mobley again asked to undergo genetic testing for evidence of MAOA deficiency. The court rejected the appeal. Mobley is now living on death row.

As our understanding of the role that genes can play in shaping behavior grows, it is only a matter of time before courts will be sufficiently impressed with evidence that a genetic defect drove a behavior that they will do exactly what Mobley's attorneys sought—permit such evidence to form the basis of an insanity defense or factor it into guidelines on sentencing. We can already see the first glimmers of such thinking. In 1990 a California judge decided not to disbar an attorney, in part because he presented evidence that the alcoholism that had made him fail his duties to his clients had a genetic basis that was beyond his control. Even more dramatic was the decision in 1994 by an Atlanta judge to release a woman from prison who was serving a life sentence for murdering her son on the grounds that she had done so under the influence of Huntington disease, a dominantly inherited, adult-onset disorder in which certain brain cells die.

On the night of July 7, 1985, Glenda Sue Caldwell walked into her 19-year-old son's bedroom and fired three shots, killing him instantly. She then went to her daughter, Susan's, room and fired at her. The bullet came so close that it burned the young woman's face. For some reason Glenda Sue did not fire again, and Susan was able to disarm her without a struggle. At the trial, Glenda Sue's lawyer argued that his client, who had lost a father and a brother to Huntington disease, almost certainly was also afflicted, and that the condition had rendered her mentally ill. At the time there was no definitive test for the disorder. As is permitted in Georgia, the jury found her guilty, but mentally ill. Judge Kenneth Kilpatrick imposed a life sentence. In prison Glenda Sue deteriorated rapidly. In 1988 she underwent brain surgery for a tumor in the frontal lobe, and in 1992 she was officially diagnosed with Huntington disease, a disorder in which cells in part of the brain called the substantia nigra die off. Although the best-known feature of the disorder, formerly called Huntington chorea, is the inability to control the movements of the limbs, some affected persons do have psychiatric problems as well.

In prison Glenda Sue suffered alone. Her marriage had ended shortly before she killed her son, and her daughter was so distraught that for two years she could not bring herself to have any contact with her mother. By the time that the diagnosis of Huntington disease was unequivocal, however, Susan had come to accept that her mother was insane the night she killed Susan's brother.

In 1992 Susan and Glenda's lawyer set out to obtain a new trial based on the fact that the doctors who made the diagnosis were willing to testify that they were certain that she was already affected with Huntington disease on the night of the killing. On August 25,1994, after a retrial without a jury, Judge Kilpatrick reversed himself and found that Glenda was not guilty by reason of insanity. Although she was no longer in the state prison, Glenda was now incarcerated in a far more terrible way. She had deteriorated so much that she was unable to leave the Georgia Regional Hospital where she had been staying before the trial. Ironically, the day after her mother was found not guilty, Susan learned from DNA testing that she too would someday develop Huntington disease. In an interview with the Atlanta *Constitution,* she was eager to talk about her predictive diagnosis. She intended, she said, to live fully and openly, capturing as many good years as she can before Huntington disease destroys her.

What if There Are Gene Variants That Predispose to Crime?

Glenda's is a tragic story, and most people would probably agree that the judge acted properly in freeing her. But what if there are persons who are genetically driven to commit heinous crimes, yet who seem by the usual methods of evaluation to be sane? Child molesters are especially likely to commit their crimes again and again. Many such individuals even express relief when they are arrested. It is possible that some of them will turn out to have a genetic abnormality that alters the manner in which certain cells in their brains respond to testosterone (the actual levels are normal). There certainly is a basis for speculating along these lines. We know that many serial sex offenders respond well to regular injections of Depo-Provera, a drug that mimics the female hormone, progesterone, and which suppresses testosterone production.

It is an extremely uncomfortable fact that criminal behavior clusters in families. According to the National Bureau of Justice, 37% of the 771,000 inmates in state prisons in 1991 have a close relative who has also been in prison. More than half of all juvenile delinquents who are imprisoned have immediate family members who have also been in prison. The more serious the crime for which a juvenile is imprisoned, the more likely he (the vast majority are teenage boys) is to have a close relative who has been in prison. Criminologists interpret these data to argue that criminality is learned, not in the streets, but at home. But it is not difficult to see how a lay person could suspect that a genetic force was at work.

Although such research is politically incorrect today, during the 1930s in Europe and the United States there were many studies of crime in families that sought evidence of a genetic influence by comparing concordance rates among groups of monozygotic (identical) twins with those among groups of dizygotic (fraternal) twins. Those studies were flawed because there were no tests to prove that twins assumed to be so were in fact identical. More problematic is that the research, especially in Germany, was done at a time when eugenic thinking was at its zenith. Allowing for these facts, the studies repeatedly found that if one monozygotic twin had been convicted of a serious crime, it was highly likely that the co-twin would have a similar history. The concordance among dizygotic twins (who share only one-quarter of their genes) was generally much lower.

In 1977 a criminologist named K. O. Christiansen reviewed the nine published studies of criminality among twins. Among them were investigations conducted in Germany (3), Holland, Finland, Japan, Norway, and the United States (2). In total, the researchers studied 216 pairs of monozygotic twins and 214 pairs of dizygotic twins. In every study the concordance rate for criminality was higher among MZ twins than among DZ twins. When he pooled the data, Christiansen found the MZ rate to be .69 and the DZ rate to be .33, strongly suggesting, but by no means proving, the influence of genetic factors. More recently (1984), sociologist David Rowe conducted a survey on delinquent behavior by sending questionnaires to virtually all twins who were in the eighth to twelfth grades in Ohio public schools. Rowe received completed questionnaires from 168 MZ and 97 same-sex DZ twin pairs (a response rate of about 50%). There was a significantly higher percentage of MZ twin pairs than DZ pairs in which both admitted delinquent behavior. Rowe concluded that the results supported a strong role for genetic factors influencing asocial behavior.

During the middle of the 20th century, a number of criminologists used adoption studies as a tool to investigate the role of genetic factors in crime. The largest effort studied all nonfamily adoptions in Denmark from 1924 to 1947. The researchers identified 14,427 adoptees and sought to study them and their biological and adoptive parents. After rigorously excluding individuals about whom there were not enough data, they still had more than 4000 male adoptees about whom they could attempt to assess parental influence on criminality. Among boys who had neither adoptive nor biological criminal parents, 13.5% had at least one criminal conviction. Where one adoptive parent had a criminal conviction, the conviction rate among the adopted boys was 14.7%. Where one biological parent had a conviction, the rate for the boys was 20%. In cases where both an adoptive and a biological parent had a conviction, the rate among the boys rose to 24.5%.

The data are more dramatic when one studies the children of recidi-vists. Parents with three or more convictions were three times more likely than noncriminal parents to have sons who, despite adoption into non-criminal families, went on to be convicted of crimes. Only 4% of the male adoptees became chronic criminals, but they were responsible for 69% of all convictions among adoptees. A similar large study in Sweden con-ducted in the early 1980s, and an American study in the 1970s, also found a strong correlation of the criminal history of biological parents with the risk of criminal conviction in children adopted away from them at birth.

Assuming there are individuals who cannot control their violent acts and who are shown to suffer from a genetic abnormality that drives them to such behavior, should they when found culpable be judged not guilty by reason of insanity and incarcerated indefinitely in a state hospital? How will new knowledge about the biologically driven behavior reshape our so-ciety's view of guilt and innocence? For the moment, these questions are purely speculative, and we have ample time to debate them. I think it likely that we will discover individuals with rare gene variants who are so driven to abnormal behavior that we will have to redefine the insanity defense to manage their disposition. Will such discoveries ultimately lead us to med-icalize the definition of crime? Someday will those who are determined to have committed certain acts be diagnosed as having a neurodevelopmen-tal disorder and then subjected to compulsory treatment rather than an incarceration? Given the fact that we have barely opened the book of the human genome and that our understanding of behavioral genetics is primitive, there is virtually no evidence to make such a prediction. Nor, of course, is there evidence upon which to reject such a future.

Discoveries like the one linking a mutation in the gene for MAOA and violent behavior pose other troubling scenarios. It would be possible to test all the young boys in each generation of the Dutch family and iden-tify those who inherited the mutation. Should the family, working with physicians, psychologists, and others, try from infancy to influence the de-velopment of those boys in the hope of countering their violent predispo-sitions? What if such an effort required the use of mind-altering drugs with serious potential side effects? When, if at all, should such drugs be given? How would we ever determine when to start? How would teachers and neighbors react to such children? How should doctors respond to the women in the family who seek prenatal diagnosis, not to avoid bearing a child with mild mental retardation, but to avoid bearing a boy who in manhood may be highly likely to attack women?

Questions like these bring us perilously close to the edges of what we wish to know about ourselves. Indeed, in regard to a possible link between genetics and violence, there are some who would prefer that the matter not be pursued at all. In 1992, after it came under attack for funding a conference of genetics and criminal behavior, the National Institutes of Health revoked a $78,000 award it had made to support a conference on that topic. This led David Wasserman, a researcher at the University of Maryland to whom the grant had been given, and his university to threaten a suit, alleging a violation of the First Amendment. After lengthy negotiations and a three-year delay, the conference was eventually held, only to be disrupted by a 1960s style sit-in. Despite the turbulence at the meeting, the scholars in attendance seemed to agree that it would be stupid to attempt genetic studies of criminal behavior, as crime is an ever-changing social construct.

As we learn more about the genetics of behavior, especially of major mental illnesses (see Chapter 9), it is very likely that we will over time develop drugs and other therapies to alter inborn predispositions. At the least, such advances will shake our faith in the 19th-century notion of "free will" that provides a cornerstone to the foundation of the criminal justice system.

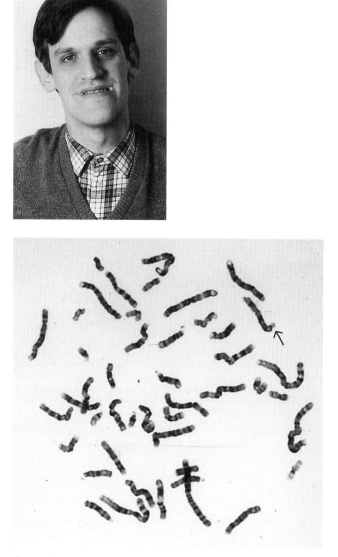

A man with Fragile X syndrome. The karyotype shows an X chromosome that appears to be broken. (Photo reprinted from Prenatal Diagnosis and Screening © 1992. Edited by D. J. H. Brock et al., p. 120, by permission of Churchill Livingstone.) (Karyotype courtesy of Genzyme Genetics.)

Wrongful Birth
What Should the Doctor Know?

Kathleen O'Brian (although this is based on an actual case, I have altered the names and places to protect the privacy of those involved) became worried that something was wrong with her son, Michael, in the winter of 1986 when he was about 16 months old. Other than a few ear infections, the little boy's health was good. He ate heartily and was big for his age, but he was just beginning to stand on his own, and he had yet to utter a word. When Mrs. O'Brian talked to her pediatrician, he was reassuring. "Each kid develops at his own pace," he said, "and it's too early to get worried." Still, he agreed to evaluate Michael's development again in three months. Michael was Bill and Kathleen O'Brian's first child, and her upbeat husband had repeatedly told her she was worrying too much about him. She went home a bit reassured, and waited for the evidence that she was wrong. It did not come.

Three months later, when Dr. Tarkington next saw Michael, their state was in the sweltering grip of a late southern spring. As Kathleen drove to the appointment, she thought about all the neighborhood children she knew who were months younger than Michael and already running about the playground while he could barely stand up on his own. She studied the pediatrician's face as he carefully performed the developmental exam, and she knew before he spoke that he now shared her concerns. "Michael is behind in his milestones," he acknowledged, "but I cannot find a reason to explain it." Together, mother and doctor again reviewed the pregnancy, her health, and family history on both sides. Everything seemed fine. The pediatrician arranged for a workup by a team of specialists.

During the next six weeks, Michael was evaluated by speech pathologists, physical therapists, social workers, occupational therapists, psychologists, audiologists, and a pediatric neurologist. His hearing was fine. Across the board, on every test, Michael, now about 20 months old, was scoring about the same as a child of 11 or 12 months. Once given a start, he could cruise along furniture, but he could not walk alone, and he was yet to speak. Again and again the experts went over the same questions. It was now obvious that something was seriously wrong, but still no one knew the cause. To the neurologist, the only hints of an underlying problem were subtle oddities in his face. Michael's head was somewhat large compared to his body, the distance between his eyes was a little more than it should be, and his palate had a higher than usual arch. Still, anyone who watched him eat an ice cream cone would see a handsome child. If he had not been developing too slowly, no doctor would have thought twice about his facial features.

Like thousands of other children each year, Michael was diagnosed with pervasive developmental disorder, sometimes called PDD, a term that at best loosely defines a child's problem but gives no hint to its cause. The clinical team arranged for an intensive program of physical therapy and speech therapy, part of a coordinated effort to maximize Michael's development, even though no one knew what was wrong. Although he realized that they would almost certainly be normal, the neurologist ordered a CT scan of Michael's head and a blood test to study his chromosomes. He also referred Michael to a clinical geneticist.

Mrs. O'Brian was by now pretty far along in another pregnancy. Very busy, hopeful that therapy would bring good improvement, she did not bring Michael to the genetics clinic. In April of 1987, she gave birth to a healthy daughter who developed normally. The following autumn, she took Michael, now 3 years old and still quite delayed, to see a geneticist.

The clinical geneticist paid particular attention to the question of whether or not Michael had unusual facial features. He agreed that Michael had a relatively large head; in addition, he noted that Michael had epicanthal folds—a little extra tissue over the inner part of his eyes (common in Asian people)—and that he had small, slightly thick ears. He also thought that Michael had somewhat loose joints, giving his limbs a large range of motion. Unlike the neurologist, he thought that the little boy's palate was normal. As is routine for examinations of children when subtle physical findings are being considered, the geneticist also examined Michael's parents. After looking at a photo of Michael's father at age 3 and examining his face during the office visit, he concluded that the slightly unusual features noted in Michael had been present in his dad during his childhood. Such familial variants are rarely of much help in explaining developmental delay. Unable to establish a diagnosis or to suggest further tests, the geneticist took a conservative approach. In his consultation note to Michael's medical chart, he advised further evaluation if and when the O'Brians decided to have another child, and he asked to see Michael again in one year.

Perhaps deciding that physicians had no more to offer them, the O'Brians did not stay in touch with the specialists. In 1991 they gave birth to a second son who soon fell behind in his development. In 1993 a different physician ordered a newly available DNA test. It revealed that Michael and his younger brother had a hereditary form of mental retardation called Fragile X syndrome. Soon after learning the diagnosis, the O'Brians sued the clinical geneticist who had examined Michael. They claimed that if the geneticist had correctly diagnosed Michael, they would have learned of the high recurrence risk they faced for having a son with the same disorder in any future pregnancy, and they would have acted to avoid that outcome. Cases like these are often called wrongful birth lawsuits.

FRAGILE X SYNDROME

The elucidation of Fragile X syndrome began in the 1940s when two doctors named Martin and Bell described a family in which some of the men had severe mental retardation and unusual physical features including a big head, simple cupped ears, a long face, and unusually large testicles. The pedigree, in which no women were affected, suggested a mutation on the X chromosome causing mental retardation. In 1969 a geneticist named Herb Lubs serendipitously noticed that when they were grown in a culture broth that was low in a vitamin called folate, the cells of a patient with an undiagnosed mental retardation contained X chromosomes that looked like they were broken. Over the ensuing decade, a number of research groups, following up on these isolated observations, made one of the most important advances in our understanding of mental retardation. In studying the cells of men living in large state facilities for the mentally retarded, they found that as many as 5–10% had X chromosomes that under the microscope looked broken when the cells had been grown in the broth without folate.

As they turned from studying institutionalized populations to other groups, scientists found that Fragile X syndrome was a common form of mental retardation. By the 1980s, we knew that about 1 in 1250 boys is born with the condition. We also knew that before they reach adolescence these boys look relatively normal; only as they leave childhood do the facial features become more striking and do the testicles tend to become unusually large. Furthermore, it became clear that the disorder is variable. Some persons have almost no features associated with the disorder and are mildly affected; other persons may be severely retarded.

In 1987, the year that Michael O'Brian was evaluated by the clinical geneticist, scientists had not yet found the gene that causes Fragile X syndrome, and no one knew why there was such a large range (from mild to severe retardation) of possible expression of the disorder. The test for Fragile X syndrome, which was quite expensive, was good, but not great. It was particularly difficult to use as a prenatal test, and most genetic labs were reluctant to offer it for that purpose. Among young boys with unexplained developmental delay, the chances were about 2–3 out of 100 that if a doctor ordered a Fragile X test the results would be positive.

During the 1990s, there was immense progress in understanding the molecular basis for this disorder. Today we know that Fragile X chromosome is caused by an unusual mutation on the long arm of the X chromosome. In affected individuals, a three-base-pair unit of DNA—CAG—that is normally present in a certain range of copy numbers has expanded far beyond its usual length. This disrupts the protein for which the gene codes. It turns out that about a dozen other neurological disorders are caused by similar stuttering mutations in other genes. Although the Fragile X gene has been cloned, we still know relatively little about the biochemical tasks that its protein performs.

Because she bore sons with the disorder, Mrs. O'Brian is a carrier of the condition. One of her two X chromosomes has a longer than normal stretch of these repeating units that expanded further during the formation of her egg cell. Unfortunately, each son arose from an egg that contained the X chromosome with the mutation. Unbeknown to her until the diagnosis was made, every time Kathleen became pregnant, she had a 1 in 4 chance of bearing a son with the disorder (a 1 in 2 chance of having a boy multiplied by a 1 in 2 chance that he had inherited her mutation-bearing chromosome).

WRONGFUL BIRTH AND WRONGFUL LIFE

In essence, Mrs. O'Brian's lawsuit argued that by failing to make the diagnosis of Fragile X syndrome in Michael, the geneticist deprived her of information that would have allowed her to avoid the pregnancy that resulted in the birth of a second child with mental retardation.

The first wrongful birth lawsuit in the United States was litigated in the late 1960s by a New Jersey woman who gave birth to a child with severe birth defects caused by the rubella virus, and who subsequently learned that her physician had diagnosed her as having a rubella infection while she was about 15 weeks pregnant. At the time, abortion was illegal in New Jersey, a fact that was crucial to the defense of the malpractice case. The obstetrician argued that he did not warn the woman about her risk of having a child with severe disabilities because the only act she could have then taken to avoid this outcome was illegal. The obstetrician ultimately won the case, despite the fact that the woman could have legally obtained an abortion in New York.

The legal climate changed greatly over the next few years. During the late 1960s and early 1970s, several leading appellate courts issued opinions concerning the dimensions of informed consent in medicine that strongly emphasized the physician's duty to disclose. In January of 1973, the United States Supreme Court handed down *Roe v. Wade*, which held that there is a constitutionally guaranteed right to privacy, which includes a woman's right to obtain an abortion. It was about then that amniocentesis to obtain fetal cells for chromosomal studies became widely available. In 1975, when it was asked to decide a case virtually identical to the New Jersey case alleging failure to warn of the consequences of a rubella infection, the Texas Supreme Court ruled that a woman could sue her physicians for failing to share important risk information.

Over the last 20 years, there have been hundreds of wrongful birth cases brought by women against physicians. Broadly speaking, they fall into one of two categories. Some, like the O'Brian case, claim genetic malpractice—that a physician failed to make a diagnosis that he or she should have made. Far more common are suits alleging that a doctor failed to warn a woman about a risk that all women (regardless of family history) like her should be told. For example, women who will be 35 at the time they deliver are at much higher risk than 25-year-old women of giving birth to a child with mental retardation due to Down syndrome. This condition, caused by an extra chromosome 21 (usually due to faulty separation of chromosomes as the egg completes a process called meiosis), is easy to diagnose if a woman undergoes amniocentesis to obtain fetal cells for study. During the late 1970s and early 1980s, high courts in a number of states held that physicians caring for pregnant women have a duty to warn them of the age-associated risk of having a fetus with Down syndrome. If a woman who gave birth to an affected child could prove to the jury that the physician had not warned her, she could prevail in a wrongful birth lawsuit that might result in a large monetary award.

The courts struggled mightily to define the dimensions of this new course of action. Judges sharply disagreed on the nature of the injury and proper scope of an award for damages. Unlike most malpractice cases, in wrongful birth cases the physician is not accused of acting or failing to act in a manner that caused an injury. The future of a child with Down syndrome, for example, is determined at the moment of conception when a sperm carrying one copy of chromosome 21 penetrates an egg that has two copies. All genetic disorders begin at conception. This fundamental biological truth caused all but a few state courts to reject lawsuits called "wrongful life" cases that were often filed in parallel with wrongful birth actions.

Wrongful life cases grow out of the same events, but unlike wrongful birth cases in which the woman is the plaintiff, the case is filed in the name of the affected child. In essence, in such cases the child plaintiff must argue that if a physician had properly warned a woman about a risk she faced in childbearing, she would have acted to avoid his or her birth. To many judges it seemed absurd to permit a lawsuit which argued that if proper medical care had been rendered to his or her mother, the plaintiff would not exist. How, judges asked rhetorically again and again, could a court compare the value of a "damaged" life (caused by a genetic disorder) to nonexistence?

The damage issue in wrongful birth cases has also troubled many judges and led state courts to craft solutions that differ substantially in scope from one another. Most courts began their analysis by asserting that a child with mental retardation or some other disorder is still a valued being who enriches the life of his or her family. In calculating damages, this notion of benefit acts as an offset to the dramatic psychological and social impact on a family of having a child with a serious disorder. In many states, judges do not permit a jury which finds that a physician failed to warn a woman about a well-characterized risk to award any damages for emotional harm. They restrict the scope of a monetary award to the special costs of raising such a child, a number that can be large, but is usually much less than the awards made in other kind of "brain-damaged baby" cases where awards for emotional damages often run into the millions.

THE LAWSUIT

The O'Brian lawsuit, which took nearly five years to get to trial, ultimately came down to just two questions. Did the clinical geneticist who examined Michael on one occasion late in 1987 violate a standard of care in failing to recommend a test for Fragile X syndrome? Should the geneticist and the other physicians involved in caring for Michael and his family have been more forceful in warning about recurrence risk? As the years slipped away and each side became more emotionally and financially involved, all prospects for a settlement vanished. When the case was called, the jury, twelve men and women plucked out of life's routine to decide an issue of high drama, sat for six weeks in a cramped courtroom facing a woman who believed that she had been horribly wronged and a genetic physician who believed that he had done a good job on a hard case and given his best advice.

As the opposing attorneys worked their way through the reams of medical records and volumes of deposition transcripts piled high on their tables, as one after another of the dozens of witnesses who had become involved testified under oath, it became steadily more clear that the outcome would hinge on how the jury regarded the conduct of the clinical geneticist during a few minutes on a December afternoon nearly ten years earlier. Had he done an adequate evaluation of Michael's problem? Should he have ordered a Fragile X test? Did he adequately warn the O'Brians about risk in future pregnancies? To decide those questions the jury had to weigh the geneticist's conduct by the standards existing then, not now. The plaintiffs called two geneticists to testify on their views as to the standard of practice concerning the genetic evaluation of a child with unexplained developmental delay in 1987. I was an expert witness on this matter for the defense.

The plaintiff's attorney, a seasoned malpractice lawyer who favored bright plaid sport coats, was demanding $20,000,000 to settle the case. He left no item unexplored, devoting two full days just to taking the testimony of his two expert witnesses. They made three important points:

(1) In their opinion, during late 1987 the standard of care for a clinical geneticist evaluating a boy with unexplained developmental delay was to routinely order a chromosomal test for Fragile X. (2) Even if that had not been the standard of care, because Michael had facial features that were suggestive of the syndrome, the geneticist should have ordered the test. (3) The physicians had not adequately warned the O'Brians that in cases where a child has a serious developmental disability of unknown cause the recurrence risk for future pregnancies could be as high as 1 in 4.

I was the first witness called by the defense on the morning of the 27th day of the trial. The plaintiff's attorney had deposed me months earlier, at which time he questioned me under oath for hours about my views on these issues. He knew exactly what the defense attorney would ask me, how I would answer, and what he had to do on cross-examination. After reviewing my credentials to establish my competence as an expert, the defense attorney wasted no time in moving the already lengthy trial along. Experts are allowed to give professional opinions. In an easygoing, relaxed manner, like a horseman sure of his mount, he elicited my opinion that in the fall of 1987 it was not yet routine to order a test to rule out Fragile X in every evaluation of a boy with serious developmental delay, that the clinical geneticist had performed a thorough, competent evaluation and decided upon a reasonable course of action, and that the written consultation note in the medical chart clearly indicated that the family had been adequately warned about risk in future pregnancies.

One key piece of my direct examination was that I had, for reasons unrelated to this case, polled several hundred clinical geneticists to determine their opinion about the routine use of Fragile X testing. The vast majority thought that for cases evaluated after 1991, the year the causative gene was cloned and a new, more accurate DNA-based test became available, clinical geneticists should, regardless of the family history and physical exam, test for Fragile X. However, a vast majority also believed that in 1987 it was not standard of care to order the test. My direct examination lasted only about two hours.

The plaintiff's attorney faced a key decision. On the basis of his opinion of the impact of my testimony on the jury and his assessment of how I would handle his hostile cross-examination, he could decide either to wave me off the witness stand or do battle with me. If he said, "No questions, your honor," I would step down, immediately leave the courtroom, and begin to fade from the jury's collective memory. As days passed and the trial continued, the impact of my views might weaken. I had not, after all, been on the stand for that long. If the jury accepted what I said as correct, the plaintiffs would lose the case. Instead of $20,000,000, the attorney would win nothing. In a few seconds, using a mental calculus to which no one else was privy, he decided to try to attack my opinion in the hope of diminishing its weight in the minds of the jurors.

For most of the next four hours, the attorney, having quite accurately identified the weak point in my opinion, hammered away at it. Was Fragile X syndrome the most common genetic form of mental retardation? Yes. Were clinical geneticists aware of that in 1987? Yes. Was the chromosomal test effective at diagnosing affected individuals? Yes. Was it readily available? Yes. Then, why doctor, was the geneticist's failure to order the test not malpractice?

Everything seems so neatly packaged in hindsight. But in a lawsuit you cannot apply today's knowledge to circumstances that arose a decade ago. Geneticists were familiar with Fragile X syndrome in 1987, but not nearly so much as in 1997. A chromosomal test was available, but it was expensive and cumbersome, and it was likely to be diagnostic in young boys with developmental delay only about 3 times in 100. Put another way, a clinical geneticist who sees relatively few patients, many of whom do not have a clear diagnosis, might order Fragile X tests for decades before getting back a positive result. In my view, the geneticist made a judgment call. He had carefully considered Michael's subtle facial features and concluded that they were familial, not that they suggested Fragile X syndrome. In retrospect, he was wrong, but the process he used in reaching his decision was right. The only way he should be held liable, I maintained, is if the standard of care at that time required that all doctors evaluating children like Michael order a Fragile X test. In my view such a standard was not in place. The jury agreed. A week after my testimony the defense attorney called me to say that after being sequestered for only a couple of hours, the jury had found for the defense.

The stakes in this case were much larger than they might appear. In the United States during the seven years from 1984 to 1991, about 10,000–15,000 boys were born with Fragile X syndrome. Most were not diagnosed until they were well along in their childhood, and in thousands of cases their parents, unaware of their genetic risk, had more children who also had a substantial risk of being affected. Girls, too, may have a mild form of the disorder (less severe because their other, normal, X offsets the one with the mutation in about half of their cells, just as it does in hemophilia or muscular dystrophy). There are thousands of families in which two children have Fragile X syndrome. If the O'Brians had won their lawsuit, hundreds of families with a similar history might also have filed wrongful birth lawsuits. The O'Brian suit would have been replayed in every state.

Every lawsuit that contests the standard of care in medicine is more than a dispute between two parties. It is a debate about social policy. At some point, it is right to decide that an expert physician to whom a couple turns for advice has erred in not ordering an easily available test, but we should first establish a rough consensus as to how to determine that the time has come to do this. The best way I know to do this is to ask the experts collectively to decide the matter among themselves and then to publish their decision and live by it. Wrongful birth lawsuits are a painful and inefficient means by which to decide social policy. By their very nature they are retrospective, and every finding of liability resonates throughout the land, stimulating many more lawsuits.

BEHAVIOR

Do Genes Make Us the Way We Are?

Emil Kraepelin. *(Photo from the National Library of Medicine.)*

Mental Illness
How Much Is Genetic?

Manic-depressive Illness

The gentle, rolling hills and beautiful farms of Lancaster County, Pennsylvania seem a most unlikely place to study the genetics of manic-depressive illness, but this lovely region is also the home of a special community of people, the Old Order Amish, a group whose way of life makes them of special interest to geneticists. For nearly 200 years, the Old Order Amish, descendants of devoutly religious, hardworking German immigrants, have farmed their land, clinging to most of their traditional practices even as modernity has surrounded them. Because all 12,000 of them are descended from about 30 persons who arrived in America in the early 1700s, and because they almost always marry within their small community, the Amish have a population profile that makes it easier to unravel the genetic component of any disorder. In addition, they have an increased prevalence of several rare genetic disorders, including a form of dwarfism, a problem that led them to Dr. Victor McKusick of Johns Hopkins University School of Medicine nearly 50 years ago. As gene mapping techniques improved, McKusick realized that they are an ideal group in which to try to map genes that predispose to complex disorders like mental illness. Mental illness is not really more common among the Amish; it is just easier to define the genetic influences.

During the early 1970s, one of McKusick's students, Dr. Janice Egeland, now a psychiatrist at the University of Miami, decided to make the huge commitment of time and energy that it would take for an outsider to win the trust of the Amish so that she could study the problem of manic-depressive illness (also called bipolar illness or bipolar disorder) in their community. Dr. Egeland's relationship with the Amish, which took nearly a decade to establish, has been ongoing for nearly 30 years.

Serious depression, of which manic depression is one form, is among the most incapacitating, costly, and misunderstood illnesses in the Western world. A 1993 study estimated that each year 11 million Americans suffer bouts of clinical depression leading them to miss 290 million days of work. It estimated that this cost American business (in days lost from work and costs of care) more than $43 billion a year, ranking it second only to heart disease. Of the 11 million persons affected with clinical depression, the study estimated that 1.8 million had the manic-depressive form of the illness. More recent estimates have put the number at 2.5 million.

Everyone has had some experience with sadness and loss and can empathize with persons caught in the grip of severe depression. Manic-depressive illness is more mysterious. During the up or manic phase, patients may go with virtually no sleep for days on end. Gripped with what seems like euphoria, they spend hours planning and trying to carry out absurdly grandiose or just plain foolish tasks. One famous autobiographical description of mania comes from the writer, Clifford Beers. While at his most manic, he felt an irresistible compulsion to write letters. Regular stationary was much too confining. Given huge rolls of paper, he would write letters that stretched the length of corridors. One measured a full 100 feet. At times his output was measured at 1800 words an hour for hours on end. This bizarre case report resonates with an experience I had as a medical student. I became involved in the care of a man in a manic phase who had spent 48 consecutive hours just prior to his admission to the hospital repeatedly taking apart and reassembling a refrigerator that was in full working order.

Manic-depressive illness is not a classic Mendelian genetic disorder, but many individuals who are incapacitated by it also have relatives who suffer from it or from unipolar depression (depression without a manic side). The Amish are not at particularly high risk for manic-depressive illness, but there are good reasons why Dr. Egeland and other researchers would want to study the heritability of this complex disorder in them. They have large families who stay physically close through the generations, they shun alcohol and drugs (the regular use of which confounds efforts to diagnose clinical depression), and they are relatively inbred. This means that if several persons in the same family develop a disorder with a meaningful genetic component, it is far more likely that the hereditary contribution is due to the same gene in each case.

In the 1930s European psychiatrists began to study the prevalence of severe depression and manic-depressive illness in the general population. Studies in several different nations were remarkably congruent. They showed that about 1 in 200 adults suffers with a severe form of this mood ·disorder. To confirm their clinical impression that relatives of patients with manic depression were at higher than average risk for becoming ill with this disorder, many of the same psychiatrists conducted family studies. The key task in family studies is to compare the risk of having the same disease among close relatives of the index case (the first person to come to them for care) to the population at large. A dozen or so major studies conducted before 1960, again mostly in Europe, were in close agreement. First-degree relatives of persons with major mood disorders were about 20 times more likely than the general population to also be affected. Put another way, about 1 out of 10 of the parents, siblings, or children of persons with unipolar or bipolar depression also was afflicted.

During the 1970s and 1980s, more methodologically rigorous studies continued to find that close relatives of persons with bipolar disease were at much increased risk for the disorder. However, those studies found less evidence than had earlier investigations that close relatives of persons with unipolar disease were at higher than average risk for developing severe depression. Overall, by 1985 the various family studies strongly suggested that manic-depressive illness has a higher genetic load than does the more common unipolar depression. One important reason for the differences between the earlier and later studies was that the diagnosis of unipolar depression has in more recent times become more rigorous. Some of the persons who might have been given this diagnosis in 1940 would not be so labeled today.

In addition to family studies, another widely used research tool in psychiatric genetics is to determine the pair-wise concordance rate of disease among twins. This complex term describes a quite simple statistic that is arrived at by assembling a study group and dividing the number of twins both of whom have a disorder by that number plus the number of twin pairs in which only one is affected. A concordance rate of 1.0 indicates that the causes of the illness are completely genetic; a concordance rate of 0 indicates that there is no detectable genetic influence. By comparing concordance rates between fraternal twins who share half their genes and identical twins, researchers derive a crude measure of the heritability of a particular phenotype. Even allowing for environmental influences, high concordance rates among identical twins are thought to suggest a strong genetic component. Recently, much energy has been expended in studying traits in identical twins reared apart, all in an effort to refute arguments that many behavioral similarities among twins are due to the impact of common familial environments. Twin studies of risk for bipolar depression, most of which were done before 1960, found a much higher concordance rate among identical twins (about 0.8) than among fraternal twins (about 0.1 to 0.2), again suggesting that manic-depressive illness has a strong genetic contribution.

Another epidemiological tool used to study the heritability of disorders is adoption studies. The usual approach is to identify a group of children placed early for adoption who were later diagnosed with a particular disorder and then compare the presence of that disease among the adoptive parents with its presence among the biological parents. The genetic hypothesis supposes that the biological (non-rearing) parents will be significantly more likely to be affected than the rearing parents. There have only been a few such studies in manic-depressive illness, and they give only lukewarm support to the genetic hypothesis.

During the early and mid-1980s, molecular biologists made great strides in locating DNA markers along the various human chromosomes. These molecular mapmakers revolutionized the search for disease genes. Clinical researchers could now investigate the presence or absence of disease in extended families and correlate the diagnosis with the presence or absence of particular DNA markers. If affected individuals across several generations virtually always have the same pair of markers, it is highly likely that the disease is strongly influenced (if not directly caused) by a gene located between those two markers. The first and easiest chromosome on which to investigate the presence or absence of disease genes is the X, because genetic disorders originating here typically affect only males. There have been many studies testing the hypothesis that some cases of manic depression are due to X-linked genes. Although the issue is not settled, overall the evidence is not particularly impressive. One Israeli study did find strong evidence of an X-linked form of the illness in five Jerusalem families.

In 1987 Dr. Egeland and her colleagues published a paper in *Nature* which offered substantial evidence that a gene predisposing to manic-depressive illness in the Amish was located in a small region of chromosome 11. The paper was based on the clinical and DNA-based study of 81 individuals in a single extended family. The psychiatrists concluded that 14 of the relatives had a definite mood disorder, including 11 with manic depression. The report, among the first efforts to scan the entire human genome to look for evidence of gene linkage, suggested that the illness was caused by a gene with a dominant mode of transmission and a moderately high degree of penetrance (chance of causing the disease to manifest). So powerful was the evidence that the scientists urged, "The demonstration by a linkage strategy that a simple genetic mechanism can account for the transmission of bipolar affective disorders in pedigree 110 should provide an impetus to analogous research on other common clinical conditions."

The report generated huge interest among scientists and the general public. The fact that two other papers published in the same issue of *Nature* did not find evidence of a gene for manic depression on chromosome 11 gave little pause. For such a complex and common disorder, it would not be surprising to find genetic heterogeneity (that changes in any one of several different genes could be causative). As predisposition to manic depression quite possibly arises due to one of several biochemical defects, it is likely that several different genes are involved. Some editorials hailed a new era in our ability to understand mental illness. On the other hand, as one put it, the news left the imagination "dangerously unfettered." Unfortunately, the scientific triumph was short-lived.

Less than two years later, further studies of the same Amish family forced the scientists to conclude that their original findings of linkage with a gene on chromosome 11 were spurious. In reanalyzing the family, the scientists (including some of the original team) extended the pedigree in two directions to add 37 individuals, diagnosed 2 more family members with bipolar disorder, and performed more detailed DNA marker studies. The crucial event was making the diagnosis of the condition in 2 persons who had DNA markers indicating they should not have the disease. The new findings now actually excluded chromosome 11 as a possible location for a predisposing gene! As Ken Kidd, a population geneticist at Yale who took part in both studies, put it for the *New York Times*, "It means we are sort of back to square one."

Twelve years have elapsed since Kidd's remark, and despite dramatic advances in gene mapping techniques, no one has yet found convincing evidence of a gene that predisposes to manic depression. Does this mean there is none? No. What it does mean is that investigating the genetics of mental illness is a devilishly difficult challenge, as will be the case for parsing the role of genes in any disorder that must often be influenced by strong environmental influences, has a wide range in age of onset and clinical manifestations, and about the diagnosis of which there is considerable disagreement even among experts. For now, we are left in the disconcerting situation that we know manic-depressive illness has in some cases an important genetic component, but we do not know what it is. In counseling families, we are reduced to quoting empirical risk figures that may or may not be relevant to them.

Ultimately, we will find and understand the genes that contribute to manic-depressive illness. One dramatic gesture made in the conviction that the genetic contribution is significant was the decision late in 1993 by the Charles A. Dana Foundation to commit $2.5 million to finding the culprit genes. The gift is being used by Johns Hopkins University, Stanford University, and Cold Spring Harbor Laboratory to conduct a painstaking search. The most important feature of the project was the task faced by the psychiatrists at Johns Hopkins of collecting 50 families in which there are at least three individuals with an unequivocal diagnosis of bipolar disorder. That took nearly four years. With this goal met, molecular biologists at Stanford began using hundreds of DNA markers to look for associations within the families of the presence or absence of disease with certain combinations of markers. The combined clinical and molecular database is stored and maintained at a laboratory in Cold Spring Harbor, New York.

Schizophrenia

One can make a fair argument that schizophrenia, not heart disease or cancer, is the most devastating illness in the western world. First well-characterized at the turn of the century by the German psychiatrists, Emil Kraepelin and Eugene Bleuler, who coined the term in 1911, the full-blown illness is marked by hallucinations, delusions, disorganized thought, negativism, and inappropriate, often flat, emotional responses. Too often forgotten today is the fact that Bleuler spoke of the disease in the plural. He recognized at least four types and speculated that heredity played an important role in each.

During the 1950s, schizophrenia accounted for more days spent in hospital in the United States than any other disease. Because of the development in the 1950s and 1960s of an array of antipsychotic drugs, hundreds of thousands of persons who in an earlier era would have spent their lives in institutions can now be cared for at home or are completely independent. Nevertheless, the disease still probably accounts for more days in hospital than any other. The background lifetime risk of developing the disorder, which typically first manifests during the teenage years, is about 1%.

Although schizophrenia typically appears as a single affected person in a family with no history of the disorder, there is a large and important subgroup of cases that are highly familial. Psychiatrists have repeatedly studied the familial clusters. They estimate that the offspring of couples in which one parent has schizophrenia have about a 10–15% risk of developing the disease. In the relatively few marriages in which both partners have the disorder, the risk to children approaches 40%. The risk to siblings of affected persons is about 5–10%. The risk for uncles and aunts, nieces and nephews, and first cousins is about half of that, nicely in keeping with a genetic hypothesis.

Twin studies have consistently found a much higher concordance rate among identical than among fraternal twin pairs. Identical twins have shown a lifetime concordance of 60–70%, whereas the lifetime risk of a fraternal twin whose co-twin is affected is about 10–15%. Of course, this is still much higher than the risk among the general population. The relatively few adoption studies of schizophrenia also generally support a genetic contribution to risk, but it is only modest.

As they had for bipolar disorder, advances in molecular biology also stimulated scientists to undertake linkage studies to look for a gene that predisposes its bearers to schizophrenia. In 1988, on the heels of the report on bipolar disorder, came a study, also published in *Nature,* asserting that there was a gene on chromosome 5 that conferred susceptibility to schizophrenia. The research was stimulated by the highly unusual case report of a Chinese man and his nephew who suffered from schizophrenia *and* had a small portion of a part of chromosome 5 stuck on to chromosome 1. One logical explanation for their schizophrenia was that the chromosomal rearrangement had damaged a gene, located at the point where the DNA had broken, which normally protected against this illness.

Focusing their mapping effort on chromosome 5, the scientists looked at five Icelandic and two British families burdened with schizophrenia across at least three generations. Psychiatrists interviewed 104 family members and diagnosed or confirmed the diagnosis in 39 persons using one standard, and replicated it in 31 of them using a more rigorous diagnostic standard. They also found many other persons who had obvious psychiatric problems, but who did not satisfy the research criteria. Using a statistic called LOD score (logarithm of the odds) analysis, the team found strong evidence that the risk of schizophrenia was compatible with having inherited a highly penetrant, dominantly acting gene on chromosome 5. They were quick to acknowledge that it was unlikely that the finding would be generalizable, as was apparent from a companion article that had found no linkage to 5 in a Swedish family.

The *Nature* paper generated wide interest among both scientists and the public and reignited the century-old debate about the relative contribution of genes and environment to the disease. The research, which was performed by a team of British scientists led by Dr. Hugh Gurling at the University of London, was sufficiently impressive that Eric Lander, one of the leading voices in gene mapping, concluded that it demonstrated "for the first time that at least some cases of schizophrenia are apparently monogenic." An editorial in *Nature* that accompanied the publication opined that "when the gene concerned has been located exactly, and perhaps its nucleotide sequence determined, it will also be possible to embark on studies of the mechanism of the disease, presumably biochemical in character, that may (with luck) be more generally applicable." Elsewhere, it described the schizophrenia in the Icelandic families as "genetically determined."

After the publication of the paper in *Nature,* many groups quickly looked for a linkage to chromosome 5 among the families they were studying. None could replicate the linkage. Less than three years later, after conducting further research with the Icelandic families, Dr. Gurling and his colleagues were forced to concede that the original findings indicating the presence of a gene for schizophrenia on chromosome 5 were erroneous. Like the story with bipolar disorder, this work shows that powerful statistical evidence of linkage within even a large pedigree fades quickly if two or three individuals are found in whom the DNA analysis does not fit with clinical diagnosis. By 1992 some scientists were arguing that the original chromosome 5 linkage report had sent the research community on a wild goose chase that had for three years slowed progress in understanding the genetics of schizophrenia.

The search continued, and in late 1993 a team studying the genetics of schizophrenia in Irish families began to alert colleagues that they had found evidence of linkage to the short arm of chromosome 6. Even before the work on the Irish families was published, other teams were rushing to replicate the findings. By this time, so much progress had been made in adding reference points to the map of the human genome that scientists had been able to develop a new approach to search for genetic linkage called two-stage scanning. This brute force technique first looks for linkage in families singled out as most likely to have a predisposing gene, such as the Icelandic families. It uses them as a template to identify areas in the genome that give hints of linkage and then seeks to replicate the findings in a completely different set of families. In 1995 the results of a two-stage genome-wide search also gave tantalizing evidence of linkage to the short arm of chromosome 6 (as well as several other areas). The paper appeared in *Nature* accompanied by three shorter articles on the same topic. Two groups reported that they had not found evidence of linkage to chromosome 6 in their families; one group reported that it had.

In the fall of 1996 at the annual meeting of the American Society of Human Genetics in San Francisco, 60 of the world's foremost researchers on the genetics of schizophrenia, many of whom had traveled thousands of miles to attend, convened a special session on chromosome 6. The results were both intriguing and frustrating. For every researcher who summarized evidence in favor of linkage, there was another who reported that his or her group had been unable to find linkage in their patients. Despite longstanding and strong evidence from family, twin, and adoption studies and an immense effort using ever more powerful mapping techniques, we have not been able to isolate a gene that predisposes to schizophrenia. Chromosome 6 still looks like a fair bet, but it will take a few years yet to find the culprit genes and establish their role.

What about the future? Have no doubt: There are genes that predispose to schizophrenia. We will map, clone, and sequence them before the year 2005. Given the huge number of people afflicted with this disorder, the news that predisposing genes have been found will stimulate the pharmaceutical industry to direct immense resources to studying the proteins coded for by these genes. Several will compete in a billion dollar research race to gain new insight into how to treat these disorders. Such investment will be amply rewarded if it results in even a single new and more effective medication.

If it turns out that mutations in one or more genes strongly predispose individuals to be at risk for schizophrenia, the use of predictive tests to assess that risk will raise immensely difficult clinical and ethical issues. Schizophrenia often begins in adolescence. Imagine this not unlikely scenario. Researchers demonstrate that in about 10% of cases of schizophrenia, almost always those in which there is a family history, patients have a mutation in a particular gene that codes for a protein that is active in brain tissue. A genetic testing company develops a DNA-based test for this mutation. Clinical research shows that persons with the mutation have about a 30% chance of developing schizophrenia before age 18, but that the other 70% of carriers do not become ill. About the same time, a pharmaceutical company develops a new medicine that proves to be moderately effective in preventing the onset of the disease if given to children who are genetically at risk, but that has unpleasant, though not life-threatening, side effects.

Should children with a family history of schizophrenia be tested? When? Will a positive result alter the way their parents treat them? If teachers learn these kids are "carriers of the schizophrenia gene," will it shape how they treat them in the classroom? Should the children at risk be given the medication that reduces the risk of developing the disorder, but has serious side effects? Is it more dangerous to the child to be labeled a carrier than to not be tested at all? This sort of scenario is likely to unfold repeatedly as we learn more about the genetics of psychiatric disease. No one has the answers to the questions it poses, but somehow we must resolve them. Hopefully, well-crafted research will show us how to combine predictive tests with information that suggests the best intervention to maximize benefit for each individual.

Great Dane and pug, illustrating wide variation within a species. *(© Jeanne White, Photo Researchers, Inc.)*

Personality
Were We Born This Way?

DOGS

My young Labrador retriever, Zoe, has an endearing (if sometimes maddening) need to be near her human family. As I write these words, she is sitting close to me, her muzzle resting on my thigh. It does no good to push her away. She needs to be this close. Zoe is affectionate to a fault. She is also a terrible beggar, willing to eat anything faintly resembling food. She will also play fetch as long as one of her humans can tolerate throwing her slimy stuffed bear across the yard.

Dog fanciers know that breeds have characteristic behaviors. The Great Dane is prized for its courage as well as its size, the tiny Chihuahua is irrepressibly lively, and the golden retriever is adored because it is adoring. Newfoundland puppies would rather swim in a bowl of water than drink it. Border collies, the product of centuries of breeding in Scotland and England, are born herders. The mere sight of sheep elicits a hypnotic stare and a low crouch. Yellow labs become clinically depressed when separated from their humans.

Yet, just a few thousand years ago, there were no dogs as we know them. If evolution is such a slow process, how could there be so many distinct breeds today? About 10,000 years ago, humans probably began to keep the cubs they found after killing wolves. Genetically programmed to belong in a group, the cubs must have accepted membership in the human clan. At first men probably used the descendants of these early canine partners mostly in hunting. As human culture matured and men became farmers, dogs, animals that possess a highly accurate ability to recognize outsiders, became valued as guards. Eventually, we came to value them for their company. In a sense, slaves became servants and, in time, friends.

Centuries before Mendel, dog breeding was already being widely practiced in many human cultures. In his great book, *The Origin of Species* (1869), Darwin cites the success of dog breeders in artificially selecting desired features in their animals to support his thesis that species evolve through natural selection. In the world of dog breeding, man is a surrogate for nature. Exerting a pressure far more intense than nature, breeders select the qualities that they want to perpetuate and permit only the exemplars to breed. Darwin was deeply interested in the success with which breeders could select for behavioral as well as physical traits. In another book, *The Descent of Man,* he speculated that although breeding had not caused dogs to gain in cunning (a trait he thought was more highly developed in the wolves from which they descended), they must have "progressed in certain moral qualities such as in affection, trustworthiness, temper, and probably in general intelligence." In Zoe's case, Darwin's speculation seems correct.

There is not necessarily a direct correlation between a highly prized physical or behavioral trait and a particular gene. But generations of breeding, often inbreeding of closely related dogs, has through trial and error made breeders quite confident of what they can expect in a litter. The reason for this is that the various breeds share many alleles (variations of particular genes) in common. One can think of breeding as an effort to identify, select for, and propagate slight improvements in phenotype that reflect small changes in the breed's gene pool about which the breeder is, albeit inchoately, knowledgeable.

Dr. Jasper Rine, a geneticist at the University of California at Berkeley, is among a handful of scientists who deserve credit for moving dog breeding from high art to hard science. Using the same techniques that permit the creation of detailed maps of the human genome, he and his colleagues started the arduous process of mapping the dog genome. Since the project began in 1990, they have mapped thousands of genetic markers more or less equally distributed across the 78 dog chromosomes. The fact that dogs have nearly twice as many chromosomes as humans does not mean that they have more genes. The number of chromosomes in each species varies widely among mammals, but the total number of genes is probably roughly the same. On each of the 78 dog chromosomes, the genes are arranged in "linkage groups" that are in many cases highly similar to the order of comparable human genes on our chromosomes.

There are several obvious reasons to study the dog genome. Of the more than 150 modern breeds, many are burdened with genetic disorders that constitute excellent models for the study of comparable human diseases. Dobermans and Scotties are at risk for hemophilia, Bedlington terriers may have abnormalities in copper metabolism akin to a rare disorder in humans called Menke's disease, Labrador retrievers and several other breeds are at high risk for congenital hip dysplasia, and beagles sometimes suffer from genetically caused seizure disorders. These are usually either X-linked (hemophilia) or autosomal recessive disorders (conditions in which the presence of a single allele in the animal is harmless, but in which the inheritance of a mutated allele from each parent causes the disorder in offspring). Unless one has a way to identify animals who carry the disease-causing recessive allele, having pups at relatively high risk for a genetic disorder is a largely unavoidable consequence of extensive inbreeding. Just as is the case with human genetic disorders, once dog geneticists have mapped the alleles that cause illness, it will be straightforward to develop tests to identify animals that carry the various recessive alleles. Breeders will then be able to avoid mating a pair of carriers, which will be a boon both to them and to people who want to purchase the animals. They will eventually be able to certify that a pup has been tested for a group of genetic disorders and has been determined to be risk-free. In 2000 Dr. Elaine Ostrander, a geneticist in Seattle who once worked with Dr. Rine and who is now the coordinator of the Dog Genome Project, reported that a group of about 15 laboratories had succeeded in correlating most of the dog genome with corresponding locations in the human genome. Along the way more than 20 canine disease genes have been mapped. Among the most important is the discovery of the gene that causes narcolepsy in Doberman pinschers, a discovery that may help understand the disease in humans.

Although eager to help in the conquest of genetic diseases among dogs, behavior geneticists seek a more elusive prize. They think that the behaviors which characterize many breeds are the product of relatively few genes, and that, once armed with a fairly informative genetic map, they will be able to track down those genes. Reversing the centuries-old strategies of breeders, they want to cross dogs of sharply different breeds that are also characterized by widely different behaviors.

Although all dog breeds are part of the same species, generations of selection through controlled breeding have almost certainly created dramatic differences in the frequency with which certain gene variants are distributed across breeds. Variation among dog breeds is, for example, much greater than is variation across human groups. Take size. Almost all adult human males weigh between 125 and 250 pounds. Men from some ethnic groups, say Northern Europeans, tend on average to be as much as 50% heavier than men from other groups, but the range is not much greater than twofold. In dogs the situation is much different. Irish wolfhounds are 50 times (5000%) heavier than Pekinese. Even so, there are no insurmountable barriers to artificially mating these two breeds and then using genetic mapping techniques to isolate genes that exert great influence on growth and adult size.

To hunt for genes that shape dog behavior, geneticists can mate animals from breeds with two sharply different behaviors, water-loving or herding or loyalty, for example, and then determine which offspring show which behaviors. They can then breed those animals and again look for the trait in offspring. In each instance, the scientists correlate the presence or absence of the phenotype in the animals from the second generation with the presence or absence of genetic markers that they know must bracket the chromosomal segment on which the gene that influences the particular behavior must reside. Over time, they can gradually narrow the stretch of DNA until they define the region that is always associated with the trait. That region will contain the gene.

If scientists succeed in finding genes that define dog behavior, they will immediately trigger two profoundly interesting questions. What are the comparable genes in humans and what are their functions? For example, once a gene associated with the herding instinct in dogs has been cloned, it may take only a matter of *minutes* to search a database of human genes to determine whether there is a human analog, which there almost certainly will be. Of course, the existence of a comparable gene in humans does not mean that it plays any discernible role in shaping human behavior, but many think that the human analog could influence some similar behavior, an intriguing and, doubtless for some, a discomforting possibility. Imagine if Darwin was correct when he suggested that in the "deep love of a dog for his master" we discern a "distant approach" to religious feelings in humans! Might there be human genes for religiosity? Fortunately, for those of us who squirm at such analogies, it will be many years before we make these kinds of connections.

BED-WETTING

We still know almost nothing about the role played by particular genes in shaping the nuances of human personality. The main obstacles to exploring the matter are the same as those that have hobbled efforts to discover genes that strongly influence risk for psychiatric disorders. Observers cannot consistently agree on the boundaries of subtle behavioral phenotypes in humans. Where are the lines that separate those who are "unusually" shy from those who are "on the shy side" or the individual who is relatively at ease among strangers from the natural extrovert? Can such boundaries ever be drawn? How are personality traits shaped by climate, cultural and religious upbringing, education, gender bias, racial discrimination, and myriad other forces? Is not a trait like introversion or extroversion heavily influenced by these and a host of other factors? Of course.

Still, we all routinely use and are comfortable with labels that capture the essence of human personality. "She's shy" or "He's the life of the party," we often say. Such phrases can even extend to ethnic groups. We speak of "dour Scots" and "talkative Italians." Deeply embedded in our language, such phrases suggest that human personality has a few distinctive features and that they are heavily influenced by genes.

Good behavioral geneticists are the first to agree that the challenges to understanding the genetic variance in personality are difficult, but thanks to our ever-growing knowledge of the human genome, they no longer view them as insurmountable. Furthermore, they are sure that a deeper understanding of the genetic components of personality and behavior will yield knowledge of great value. To open a discussion about genes and personality, I start with a common behavioral problem of childhood— bed-wetting—that has long been thought to be correlated with certain environmental stresses, but about which there has recently emerged convincing evidence that it is strongly influenced by genes.

Millions of children (considerably more boys than girls) persistently wet the bed after age seven. For much of this century, primary enuresis, as the child psychiatrists call it, was thought to be due to emotional problems, in many instances serious parent-child conflicts. Many psychiatrists thought that the problem arose out of conflict between a domineering parent and a passive-aggressive child. Families spent long hours with pediatricians and psychiatrists, and both parents and children felt acute embarrassment about bed-wetting, often creating tensions that harmed the daytime routine.

The first studies to suggest that bed-wetting might be genetically influenced appeared about 20 years ago. From responses to questionnaires administered to the parents and grandparents of children who were chronic bed-wetters, it became clear that in many instances the problem was highly familial. One large study showed that in families in which one parent had a problem in childhood with bed-wetting, nearly half his or her children had the problem; when both parents had been bed-wetters, about three-fourths of the children were affected. These results were compatible with the effects of a highly penetrant dominant gene.

In 1995 a group of Danish scientists uncovered powerful evidence that a variation in a gene on the long arm of chromosome 13 caused a substantial portion of serious bed-wetting. In 1973 some of their farsighted colleagues had collected and frozen white blood cells from members of 832 "normal" families, who were special only because each included at least four children, a fact that greatly helps genetic analysis. Twenty years later the researchers who sought to do gene mapping studies in families burdened by primary enuresis sent questionnaires to 655 of these families, asking whether bed-wetting was a problem. Eleven families who acknowledged a strong family history agreed to participate. The researchers compared the DNA markers of the parents who had been bed-wetters to the markers of their affected and unaffected children. All but one of the 23 parents who reported having been bed-wetters transmitted the copy of their chromosome 13 with suspect markers to the affected children, and they passed on the other copy of 13 to their unaffected children. This strongly suggests the action of a dominant gene.

How could a change in a single gene cause a problem like bed-wetting? One of several possibilities is that it could alter the function of a protein that the brain uses to tell the kidney to produce less urine at night. Children who are not as able to concentrate their urine will be unable to avoid urinating for an 8–10-hour stretch. Because kids sleep more soundly than do adults, they will be less likely to awake to avoid wetting the bed.

The genetic hypothesis has greatly changed how parents and health care professionals view bed-wetting. It has been reconceptualized as a physiological disorder that is often caused by inadequate levels of antidiuretic hormone. Child psychiatrists now focus on reassuring children and parents that bed-wetting is not a form of acting out, a sign of an anxious temperament, or the result of a personality conflict. Rather than looking for evidence of familial dysfunction, they help families cope with the disorder to reduce the risk that they will become dysfunctional. The discovery of a genetic basis for bed-wetting is impressive, but it falls far short of convincing anyone that human personality is deeply shaped by discernible genetic factors.

HAPPINESS

What fundamental question might we ask to investigate the role of genes in human personality, and how might we pursue it? Of the many ways in which we can define our own personalities and those of our relatives and friends, none is more basic than our sense about an individual's baseline state of happiness or sadness. Everyone knows people who seem upbeat or happy; most of use know others who strike us as generally downbeat or sad. Could something so fundamental as a predisposition to happiness or sadness be genetically driven? Is happiness heritable? Behavioral geneticists are already probing this question, and a few are already arguing forcefully that the answer is yes.

Among the most interesting and controversial findings are those of David Lykken and Auke Tellegen, psychologists at the University of Minnesota. In studying the heritability of mood, they pursued the time-honored technique of comparing identical with fraternal twin pairs. Their subjects were 1380 pairs of twins born in Minnesota between 1936 and 1955. They assessed "happiness" by using the Well Being scale of the Multidimensional Personality Questionnaire, a self-report inventory that asks respondents to indicate whether they agree or disagree with statements such as "I am just naturally cheerful." They found that identical twins were far more likely than fraternal twins to provide similar answers to similar questions. The cross-twin concordance for identical twins was 0.44, while for fraternal twins it was only 0.08. That is, adult identical twins were far more likely than fraternal twins to record similar answers, a difference strongly suggesting (but by no means proving) that genes have a major influence in shaping one's baseline mood.

Lykken and Tellegen followed the twins for five to ten years and then asked some of them to complete the Well Being scale a second time. This allowed them to compare the responses of each twin at two points in time and, in addition, to compare the responses of one twin at one time to the co-twin at another time. The cross-twin cross-time results among identical twins was about 0.4, much higher than the score of 0.07 recorded by fraternal twins on the same comparison. In an editorial assessing the research, Dean Hamer of the National Institutes of Health, a scientist who does similar work, suggested that the data indicated that the broad heritability of a sense of well-being is 40 to 50%. The implications of attributing such a large fraction of personality to genes are immense. It suggests, for example, that parenting style, economic status, and education have relatively little impact on the child's sense of well-being. Hamer opined that a strong genetic influence on well-being could explain other research indicating that positive life events and socioeconomic success have little impact on subjective feelings about mood. For example, lottery winners are not much happier after winning the jackpot, and people paralyzed by spinal cord injuries are (in the long run) not much sadder than average. Perhaps the most provocative aspect of the research was its suggestion that the best predictor of whether one will be happy 10 years hence is how one feels today, in essence an argument that future happiness is genetically influenced!

We have not found any happiness genes yet, but we may. In the meantime, if happiness is largely a matter of genes, why do we strive so hard to feel better than we naturally do? Lykken and Tellegen have speculated that, "It may be that trying to be happier is as futile as trying to be taller and therefore is counterproductive." A more hopeful answer is probably that we all have a range of moods, and that it might be possible for one to work to maximize the happiest possible frame of mind.

The Minnesota happiness study is only a beginning, a statistical observation based on analyzing a pile of questionnaires. Is there any evidence of particular genes that can significantly affect personality? Doing linkage studies on inbred strains of mice, some scientists have described a phenotype called "emotionality" and have traced its expression to the effects of just three genes. No such work has been done in humans, but two scientific papers have made (as yet unverified) claims that variants in a single gene can determine aspects of personality such as "novelty-seeking" and the ease with which one makes friends.

Novelty-seeking

The work on the genetics of novelty-seeking also began with a questionnaire. Robert Cloninger, a leading behavioral geneticist at Washington University, created the tridimensional personality questionnaire (TPQ), which is intended to assess four broad categories of temperament that he calls novelty-seeking, harm-avoidance, reward-dependence, and persistence. He did this to pursue his theory that differences in temperament among people are strongly influenced by variations in the production or processing of a neurotransmitter (a chemical messenger that sends signals from one brain cell to another) called dopamine.

In 1996 some Israeli scientists reported that among 124 unrelated individuals, those subjects whose responses to the TPQ placed them high on the novelty-seeking scale were also much more likely than expected to have a particular version of the dopamine D4 receptor gene. The finding was in keeping with other knowledge about dopamine: (1) In animal studies, dopamine levels have been shown to affect exploratory behavior; (2) variations in dopamine metabolism affect how people respond to cocaine; (3) humans vary widely in how their brain cells take up the neurotransmitter; (4) the dopamine D4 receptor is found on a more restricted set of brain cells than other dopamine receptors and may be the site of action of a drug called clozapine. The authors posited that a gene variant that drove novelty-seeking might be more efficient in bringing dopamine into brain cells than is the common version.

The report on novelty-seeking was all the more impressive because it was accompanied by findings from another group that, working independently of the first, had found essentially the same results. Using a different questionnaire, the second team correlated the responses of 315 people with whether or not they had the version of the D4 receptor that the first research group had associated with novelty-seeking. The "extroversion" scores were significantly higher among those who had the same gene variant that the other researchers had associated with novelty-seeking. This was especially interesting because the second study population was quite different from the first. Almost all (95%) of the group was male and about half were gay. This is because the research was done in a laboratory that is studying the genetics of homosexuality (see Chapter 12), and the blood samples and questionnaire results were readily available. The results could

be confounded if gay males turned out to show more novelty-seeking behavior than other people. The studies did not make grand claims. The data
suggested that the D4 receptor accounts for about 10% of the genetic variation in novelty-seeking. Yet, it is a breakthrough of sorts. Together, these
papers (however modest) constitute the first replicated association in humans of a specific genetic locus with a personality trait.

SOCIABILITY

The first suggestion that there is a gene that directly influences one's sociability grew out of a study of Turner syndrome, a condition affecting about
1 in 2500 girls that is caused by being born with one (instead of the normal two) X chromosome. No boys are born without an X chromosome because for them this is fatal early in development. The lack of a second X is
often lethal to the female fetus. No one understands why many female fetuses with Turner syndrome die while many others are born and remain
healthy.

Girls with Turner syndrome can have a variety of abnormalities, the
most obvious of which is extra skin on the sides of the neck (a consequence of abnormalities in the development of the lymph system) that
give it a webbed look. However, most affected infants look normal and are
often not diagnosed until much later. The girls do grow slowly, and one of
the first hints that something is wrong is that they are unusually short. In
the not-so-distant past, one leading human genetics textbook quaintly
characterized them as "dainty little girls with pleasant personalities who
are well behaved and industrious." Unless they are treated with a drug
called oxandrolone or genetically engineered human growth hormone,
girls with Turner syndrome usually reach no more than 4 feet, 10 inches in
height. The diagnosis is usually made when the girls reach their teenage
years and do not menstruate or develop breasts. They have only rudimentary or "streak" ovaries.

Contrary to the textbook description, many girls with Turner syndrome have problems in school. Although a few have completed college
and taken on challenging jobs, many make it only partway through a standard course of public education. In one European survey of 126 women
with Turner syndrome, only 12 had completed high school, and 21 reported having been schooled in classes for children with special needs.

About one-third of the girls score in the mildly retarded range on IQ tests. Scientists have long been fascinated by the fact that there is a sharp discrepancy in various sub scores of that test. Persons with Turner syndrome tend to be unusually weak in math and to have a very poor sense of direction. They often report that they cannot read maps and have trouble finding items in their homes. As they grow up, some persons with Turner syndrome also have great trouble in social relationships.

Because they have only one X chromosome, girls with Turner syndrome represent an intriguing experiment of nature that scientists have used to explore a still poorly understood phenomenon called imprinting. There is good evidence that the action of particular genes in an individual sometimes differs sharply depending on which parent provided the chromosome. Depending on whether they are transmitted through egg or sperm, some genes may never even be activated. Through DNA marker studies, we know that about 70% of the girls with Turner syndrome have inherited their X chromosome from their mothers and about 30% from their fathers. (This is as expected, because females have two X chromosomes and males have only one.) This provides researchers with a way to ask whether genes on the X chromosome show evidence of imprinting. Do they act differently depending on parent of origin?

Aware of the wide range in social skills among girls and women with Turner syndrome, Dr. David Skuse, a behavioral scientist at the Institute of Child Health in London, and several colleagues undertook a project to determine whether affected girls could be divided into distinct groups in which their level of sociability could be correlated with whether they had inherited a maternal *or* a paternal X chromosome. After obtaining approval from local ethics committees, the group started their work by identifying persons with Turner syndrome from a national registry in England and sending questionnaires to them. Parents, school teachers, and, if they were 11 or older, the girls themselves, were each asked to respond to different survey instruments. Each respondent was asked to evaluate or self-evaluate how the particular person with Turner syndrome was doing in school and in the more general category of social adjustment. Analysis of the responses from parents (who were surveyed twice about two years apart), teachers, and the girls allowed the researchers to obtain differing perspectives on the same set of issues. The team also performed DNA studies to determine from which parent the sole X chromosome derived.

On every social measure the group of 25 girls whose X chromosome derived from their fathers scored much better than did the group of 55 girls whose X chromosome came from their mothers. Their parents, their teachers, and the 25 girls themselves all gave responses indicating that they were well adjusted. The girls with paternal X chromosomes also had better verbal skills and demonstrated much superior "higher-order executive function skills," measures that may reflect how well one will get on in everyday interactions. Only 4 of the 25 girls whose X chromosome came from their fathers needed to be in special education classes, while 22 of the 55 girls with the maternal X were in special education programs (statistically, a difference of great significance). Also intriguing was that 21 of the 29 girls over age 11 with the maternal X had significant behavioral problems, whereas only 4 of 14 similarly aged girls with the paternal X had such difficulties.

The researchers also devised a social cognition questionnaire that they administered to normal boys and girls as well as those with Turner syndrome. On this scale (in which a *higher* score indicates *more* evidence of social adjustment problems), the Turner girls with the maternal X had much higher scores than did those with the paternal X, and normal boys had much higher scores than did normal girls. The boys scored about the same as did the Turner girls with the paternal X.

From these findings the researchers concluded that they had uncovered the first solid evidence for the existence of an imprinted gene on the human X chromosome, and that it exerts great influence on the development of social skills. They argue that the gene is only expressed if it comes from the father. Put another way, it will be inactive if it is transmitted through the mother. Because normal girls get an X chromosome from each parent, the genes at this locus will be active. Boys, however, only inherit a single X and always from their mothers. Thus, they might routinely inherit a gene that may have an important role in social cognition, but which never is working to their benefit. In the (somewhat overstated) words of the researchers, this could "explain why males are markedly more vulnerable than females to pervasive developmental disorders affecting social adjustment and language, such as autism." Of interest, the investigators did find three Turner girls who carried the diagnosis of autism. All three had inherited the maternal X! In interviews with the press, Dr. Skuse described the girls with a maternal X as *not* good at "reading" body language, tone of voice, and other subtle messages. He also suggested that normal girls are "hard-wired" to pick up social skills almost instinctively, while normal boys have to work to acquire those skills.

Science journalists hailed the report as evidence that there is a gene that makes girls more adept than boys in learning social cues. But it is a giant leap to use the results of studies that combine profiles of human personalities with genetic analysis to posit that a single gene explains why boys are at higher risk for autism or have more social adjustment problems than do girls. To those who are troubled by the impact of genetic information on society, the determinism implicit in such speculative leaps seems irresponsible and scary. We will ultimately understand how genes interact with the environment to shape personality. But, our genes are, quite literally, only what we start with. Much good will come of a deeper understanding of the role of genes in the development of personality, but society will not benefit from this knowledge if we overstate its value or tolerate its misuse.

Johann Sebastian Bach and sons, by Balthasar Denner, c. 1730. (Photo: AKG London.)

Talent

Nature or Nurture?

PERFECT PITCH

A musical tone is defined by the speed with which a vibrating instrument generates a sound wave. When an orchestra tunes, the players key to the first violin's A, which propagates across the stage at 440 cycles per second. When Dr. Joseph Profita, a psychiatrist who works at the Veterans Administration Hospital in Sepulveda, California, hears a tone, he instantly recognizes it. He and about 1 in 2000 other people, including Nat King Cole, Andre Previn, and about 10% of the students at the Juilliard School of Music, were born with the capacity for perfect pitch. With appropriate early training, they hear notes as easily as you see colors. Just as you know instantly and unequivocally when you see green or purple, people with perfect pitch immediately recognize an F or a G. Dr. Profita realized that he could identify tones long before he understood anything about music.

Perfect pitch is neither necessary nor sufficient for a successful music career, but it is clearly of great value to a professional musician. Yet, despite how proficient pianists or violinists might be at their instruments, if they were not born with perfect pitch, they cannot develop it. They can with effort teach themselves to have perfect relative pitch—the ability to recognize a second tone from a reference note—but they must always have the anchor. To envious fellow musicians, their colleagues with perfect pitch must seem to come and go freely in a tonal space that they cannot enter.

Dr. Profita, who is a fine amateur pianist, has also become an amateur geneticist. For many years he has sought out other people with perfect pitch and asked whether it was shared by any of their relatives. As the anecdotal evidence that this was a heritable trait mounted, he became more interested in testing that hypothesis. He realized that he could not do so until he had a way to assess the presence or absence of perfect pitch in people who had no special interest in music. Because most people with perfect pitch are not professional musicians, to pursue the question of heritability only among musical families would severely bias the investigation. Some years ago, he and several colleagues developed a test for perfect pitch that one can administer to persons even if they cannot read music or play an instrument. By the late 1980s, he had identified 60 families in which some members had perfect pitch. From analyzing the family trees, he showed that the data on who did or did not have perfect pitch were strongly consistent with (but did not prove) the presence of a highly penetrant autosomal dominant gene. In a nutshell, about one-half of the children of persons with perfect pitch also have the trait.

In 1997 a scientific team led by Nelson Freimer at the University of California in San Francisco took research on the origins of perfect pitch to a new level. Working with Shai Shaham, a gifted musician turned geneticist, and Siamak Baharloo, he and others developed a survey instrument with which to screen individuals to ascertain the likelihood that they have perfect pitch. Shaham, incidentally, is the brother of world-renowned violinist, Gil Shaham, and of the talented concert pianist, Orli Shaham. All three children and their father have perfect pitch. Among more than 600 professional musicians, the researchers found that about 15% have perfect pitch. Interestingly, virtually all of them who do started formal music training before the age of six. This suggests that the capacity for perfect pitch can be lost if not nurtured. The research team is among the first to recruit subjects over the Internet. They are trying to recruit 100 extended families with multiple members who have perfect pitch. They hope to collect DNA from all of them and within each family compare the DNA markers among those who have the trait to those who do not have it.

What is it like to have perfect pitch? Many musicians who have been interviewed say that they recognized and had distinct feelings about notes when they were as young as three. Some experience synesthesia; they associate certain notes with definite hues such as soft rose with a D and sharp yellow with an F. The associations differ with each person. Does perfect pitch make one a better musician? Opinions vary. Some musicians wryly insist that it merely makes one much more aware of one's mistakes.

Research into the genetics of perfect pitch is unlikely to be well funded, but we will find the relevant gene soon enough. By 2002 the consensus sequence—each of the 3,000,000,000 or more DNA letters code for all the genes in the human genome—will be literally available to everyone. It will be public information, any part of which can be accessed through a National Institutes of Health website.

Many environmental factors influence the role of music in one's life, and most persons with perfect pitch are not musicians. To those who may have little or no interest in music, this "talent" could be unimportant or even a hindrance. What most of us tolerate as the dissonance of everyday life (car horns, subway wheels, pots clanging) must be particularly irritating when the deviant notes are so obvious. They might take issue with my suggestion that perfect pitch is an example of a talent (defined in my tattered Webster's as a "natural endowment"). Still, I think it a fair example of a trait that some of us (my tone-deaf self included) would love to have, but realize is forever out of reach.

GALTON'S LEGACY

Talent is ineffable and elusive. Although there is a good chance that no two people will agree on what it is or who has it and who does not, or who among people with particular talents is more talented, achieving rough consensus about who occupies the upper reaches in some firmament is not too hard. Michael Jordan, Tiger Woods, Stefi Graf, Leontyne Price, Seiji Ozawa, and Yo-Yo Ma have each reached greatness in his or her field. How did they do it? What is the nature of the complex interaction of genes with environment that has propelled them to such heights? What, if anything, can we say about their genes? We really have no idea what it is that enables a person to do so extraordinarily well at some endeavor that he or she escapes from the crowd of the very good to join the inner circle of the great. Even though there have been extraordinary advances in human genetics, we are not remotely close to being able to address questions such as what role a composer's particular genetic profile played in contributing to his or her success.

Efforts to parse the hereditary contribution to talent, or at least to attainments readily acknowledged and admired by others, began with a serious misstep. Francis Galton, a Victorian polymath and a cousin of Charles Darwin, was among the first to study the relationship between natural endowments and success in life. In 1865 he published two articles in *MacMillan's* magazine on the accomplishments of relatives of then-prominent English judges during the prior two centuries. He reported that an impressive number of relatives had also achieved eminence and posited that they were members of what today we might call a genetic elite. In 1869 he published a book, *Hereditary Genius,* with the highly presumptuous subtitle, "An Inquiry into its Laws and Consequences." For each field in which he was interested—the work of judges, statesmen, military leaders, writers, scientists, musicians, painters, clergy, oarsmen, and wrestlers—Galton identified those men in England whom he thought most eminent. He then investigated whether or not they had eminent relatives and, if so, whether the number was much in excess of that which would be attributed to chance. In all the fields he investigated, Galton found substantial evidence that achievement was strongly influenced by heredity.

By today's research standards, his work is badly flawed. He barely considered, indeed virtually dismissed, the contribution of wealth, schooling, and social class to success in life. In 19th century England there were few paths to success for even the most talented person born into the poverty in which the mass of people lived. But, in providing what passed as scientific proof that inborn talents had allowed individuals to emerge over time as founders of family dynasties that continued through the generations to dominate English politics, arts, and science, Galton wrote for receptive readers. No one criticized him for failing to acknowledge that social class locked talented people out of the niches in which their abilities might flourish. A social Darwinian, Galton coldly argued that if a person had talent worth worrying about, that individual would break through to success.

Although Galton is correctly credited with being the founder of the eugenics movement (see Chapter 24), for most of the period when it flourished (about 1900–1940), the major focus was on preventing the flow of "defective" genes, not encouraging the flow of "positive" genes. Galton and his English disciples were among the few who instead sought to encourage the scions of talented families to pick talented spouses. In the United States, Teddy Roosevelt was among those ardent positive eugenicists who boldly urged healthy, intelligent young Americans to marry and have big families, the best antidote, they thought, to the polluting of the American "gene pool" caused by immigrants from southern and eastern Europe. Roosevelt's exhortations presaged by four decades the corrupt Nazi version of his vision known as Lebensborn, a determined effort to match young persons of good Teutonic stock to become parents for the father-land.

FAMOUS FAMILIES

When we consider the extraordinary accomplishments of groups of persons in particular families, the possibility that genes are a driving force in talent is not easy to dismiss. Who does not know that several of the greatest composers were raised by parents who were accomplished musicians? The Bach family includes at least 20 musicians across four generations who demonstrated special talent. Other great composers evidenced striking talent at such precocious ages that it is hard to imagine it as a product of great teaching. Mozart, Beethoven, Mendelssohn, and Schubert were publishing scores as children or young adolescents. In painting, Ruysdael and Titian (to name just two of many) were born into families of artists and both showed exceptional talent in childhood. Such family histories stimulated Charles Davenport, one of the founders of human genetics in the United States, to speculate naively that great artistic talent arose in children only if they inherited a pair of predisposing recessive genes!

By today's standards, Galton's research methods would fail peer review, but he still casts a long shadow. Throughout the 20th century, our culture has continued to be fascinated by families in which extraordinary talent, abilities not easily attributable solely to economic security or social class, persists through the generations. Consider, for example, an article in the *New York Times Magazine* entitled, "Natasha Richardson and the Redgrave Dynasty." The author began his story on the dynamic young actress, the fourth generation in the family to win fame on stage, in this way: "Her mother's daughter, her aunt's niece, her grandfather's granddaughter, her sister's sister—the career of the Americanized actress invites the intriguing question: Is talent hereditary?"

Natasha Richardson is the daughter of Vanessa Redgrave who vaulted to fame in 1967 with her performance in the film *Blow Up*. Vanessa is the daughter of Rachel Kempson and Sir Michael Redgrave, both acclaimed British actors. On the evening when Vanessa was born in 1937, Sir Michael was playing Laertes in *Hamlet*. Lawrence Olivier, who was playing the title role, stepped to the footlights and announced her birth to the audience, predicting that she would be a great actress—which she became. Vanessa's daughter, Natasha, appeared in her first film at age 4, and by 18 she was acting in London's West End. The author describes her as having "the family aptitude for emotional extremity." Elsewhere he says, "She has inherited her mother's smoky voice and her genius for emotional revelation, yet she is determined to make her career independently."

Which is it—heredity or environment? Natasha grew up in a wealthy household that was at the epicenter of theatrical life in England. As the author put it, she "inevitably attended drama school" and was pushed into theater work while still a young child. Throughout her childhood she watched her grandparents and mother win acclaim on stage. Although she is no doubt hurt by allegations of nepotism, Natasha Richardson would also have to acknowledge that many more theater doors were open to her than to her unknown competitors. The answer, of course, to the question is "both." Natasha may well have inherited vast complexes of as-yet-undescribed genes that fashioned her voice or cheekbones in ways that recall her mother's. On the other hand, if her mother had rejected the theater and made every effort to suppress an urge to act in her daughter, Natasha might well have grown up to be a lawyer or a teacher.

In 1997, four years after running the article on the Redgrave family, the *New York Times* again took up the heritability of talent in a piece entitled, "When Creativity Runs In the Family." The journalist, Loch Adamson, opened: "Artists may well be born, but they are also made." He opined, "it is tempting to ascribe artistic talent, even creativity, to genetic or neurological factors. Yet, even art and creativity are products of nature. They must all be nurtured, recognized, and supported." The article reviewed the impact upon talented children—Julian Fleisher, a jazz singer; Betye and Lezley Saar, visual artists; and Hallie Foote, an actress—of growing up under the influence of an extremely gifted and successful parent. Each child wound up working in the same artistic field as the parent, but all noted that observing the creative struggles (and, sometime, the despair) of their parents had at times dissuaded them from the road they ultimately decided to travel. Articles like this fail to do the obvious, which is to remind us of the far greater numbers of families in which a child does not follow a parent into artistic greatness. It seems to me that Ruth Richards, a psychiatrist at the Saybrook Institute in San Francisco, who was interviewed for the article, got it just right. "With the right combination of elements—

psychological, social, and biological—running in families, there is a real possibility of sparking eminent creativity."

Richards reminds us to avoid genetic reductionism (the notion that a person's destiny is encoded in his or her DNA). The shameful history of the eugenics movement (Chapter 24) proves that genetic reductionism flourished for a good chunk of the 20th century. It is, thus, difficult to dismiss concerns that it could rise again. During the 1990s, genetic imagery became a favorite of the advertising world and insinuated itself into every corner of modern life. Among the many early favorites I collected is a two-page glossy ad for the British Sterling automobile that graced the inside cover of *Sports Illustrated* in 1990 under the lead, "The remarkable handling of a Sterling. It's in our genes." Beneath, a beautiful photo of the automobile parked majestically before a brick mansion ran the explanation for its quiet, comfortable ride. "It's British. Nobody understands suspension quite like we do."

INTELLIGENCE

For most of this century, the center of the controversy over the genetic contribution to talent has been the debate over the heritability of intelligence. Argument perennially swirls around three issues. Can intelligence be defined? Can it be measured? To what extent is the capacity for intelligent behavior inherited? In 1906 the psychologist Henry Herbert Goddard imported the IQ test from France, where it had been developed by Alfred Binet and Theodore Simon as part of their struggle to determine the learning capacity of children with mental retardation. The test was quickly adopted by American psychologists. Just a few years later, government officials decided to administer it to every man who enlisted in the U.S. military. For most of the 20th century, taking an IQ test was almost a rite of passage for American children. Only recently has its validity been sufficiently challenged to cause doubts as to its value in guiding educational choices for children.

Is there a readily measurable human trait that is positively associated with achieving economic and social success in adult life and that is far more influenced by genes than environment? Unquestionably. It is height. Many studies have shown that most of the variance in adult height is attributable to genes. A number of impressive efforts compared final adult height among identical twins reared apart with that of identical twins, nonidentical twins, and siblings reared together. For example, in their 1937 book, *Twins: A Study of Heredity and Environment,* Newman, Freeman, and Holzinger, three scientists at the University of Chicago, studied height in 50 pairs of identical and nonidentical twins reared together, 52 pairs of siblings reared together, and 19 pairs of identical twins reared apart. More than 70% of the identical twin pairs, whether reared together or apart, had a final adult height that differed by less than 1.9 centimeters (three-quarters of an inch), whereas only 30% of nonidentical twins and sib pairs reared together (that is, in the same general environment) were so similar. The correlation coefficients for height between pairs of identical twins reared together or apart were much higher (greater than 0.9 where 1.0 would indicate that there was no discernible evidence of any environmental influence) than for nonidentical twins and sibs born at different times, suggesting that environmental factors played only a small role. Across all groups, the correlation coefficients were strongly positive, suggesting a major role for genes in determining height.

More than a few studies show that tall people are more likely to achieve economic and political success in American culture than are short people. Who does not know the trivial fact that all but a few American presidents have been significantly above average height? Although those of us, like myself, who are on the short side might not like to hear this, it is true. However, I suspect that few comparatively short people (those within 1–2 standard deviations from the mean) lose any sleep over their height after adolescence. This is not so much because they know that they can't change their height as because they realize that such a statistic has no predictive value for any one individual. Put another way, they know that there is substantial scatter in the correlation between height and success. Many short people do much better in life than do many tall people. Whatever the height advantage is, in the grand scheme of things, it is relatively small.

Whether intelligence, like height, is easily measured, highly heritable, and significantly correlated with socioeconomic success has been frequently and bitterly debated. During the 1920s, 1930s, and 1940s, due mainly to the influence of several large longitudinal studies, especially that of 1500 gifted (all but a few of whom had an IQ above 140) children in California by Stanford psychologist Lewis Terman, the dominant view embraced by American academia was that intelligence is measurable and correlates well with later success. Terman did not focus on heritability, but he was struck by the observation that siblings of his gifted children (who were nicknamed "Termites") were also often gifted. During the early 1930s, Terman was convinced that the persistent differences among racial groups in mean IQ score could not be accounted for by environmental differences and, thus, must have a genetic basis, but as time went by, he backed away from this view.

Although children who take an IQ test several times over several years tend to receive quite similar scores, and although there is much evidence to confirm Terman's observation that persons who score high tend to enjoy socioeconomic success later in life, for the last 40 years there has been intractable debate as to what the test really measures and as to whether it is culturally biased. Some fascinating aspects of this debate are succinctly told by Stephen Jay Gould in *The Mismeasure of Man,* which won the award from the National Book Critics Circle for the best nonfiction book published in 1981.

Since the late 1960s, the influence and use of IQ testing in American society has been waning, in part because of accusations that it is biased in favor of white middle class children. One of the major inflection points in this trend occurred in 1968 when Arthur Jensen, an educational psychologist at Berkeley, reported that early intervention programs intended to help disadvantaged, predominantly black, youth were not providing any discernible impact on school performance. He argued that the 15-point difference in mean IQ scores between whites and blacks reflected an underlying biology that special educational efforts could not overcome. A few prominent psychologists, such as Hans Eysenck in Britain, supported Jensen, but the vast majority of educators rejected his work. Appearing at the zenith of our society's interest in affirmative action programs, his arguments also drew a fusillade of angry responses from less well informed critics.

Decades of research have consistently showed that as a group, African-Americans on average score consistently lower than, and Asian-Americans consistently score higher than, do whites. Of course, this is merely a summary statement about the results of testing in large populations that when compared show a huge overlap of test scores. Within these groups there are many black children who score higher than do many white children. Statistical observations about groups are irrelevant to the test score obtained for any individual child. Furthermore, even if IQ testing is a reliable indicator of the capacity for intelligence, the test results suggest nothing about why persons who score high are intelligent.

The revolution in molecular genetics will soon permit us to make a serious attempt at answering that question. Of the 40,000 or so genes that are active at some time or another in human brain cells, it is possible that a much smaller number (perhaps 10 or 20 or 100) play a special role in the development of what we call intelligence. Possibilities include genes that code for neurotransmitters or the receptors with which they interact, genes that control the manner in which nerve cells migrate to their appropriate location in the developing brain, and genes that determine the manner in which nerves become insulated, thus affecting the speed with which information is propagated along them. It is highly likely that scientists will elucidate such genes and, eventually, understand their role in neural physiology.

A few intrepid souls, such as behavioral geneticist Robert Plomin, are already trying. He is using a technique called quantitative trait loci (QTL) mapping to attempt to find alleles that are common in people who have quite high IQ scores, and are less common in persons with quite low IQ scores. Among his most fascinating studies is work he is performing in collaboration with educational psychologists at Vanderbilt University. Since 1998 Peabody College at Vanderbilt has been the home of a long-term project called the Study of Mathematically Precocious Youth (SMPY), often simply called the Stanley Study, in honor of its originator, Julian C. Stanley, who began to study mathematically gifted children at Johns Hopkins University in 1971. Over nearly three decades, the SMPY has collected tens of thousands of children who record dramatically high scores on IQ tests. In 1998 Plomin began to collect DNA samples from a special cohort of children whose scores meet a selection criteria that is only expected in 1 of every 30,000 randomly ascertained kids. He is convinced that his cohort will provide the world's first DNA marker for general intelligence. I doubt he will succeed. If even as few as 100 genes control most of one's potential to do well on an IQ test, each is likely to make only a small contribution (on the order of 1%). It is extremely unlikely that Dr. Plomin will ever be able to perform a large enough study to isolate these effects from the general background and study them further.

Plomin might do far better searching for genes that "cause" intelligence by first seeking genes that cause mental retardation, a topic on which we have, in some cases, a slightly better grip. We know quite a few genes that must be crucial to intelligence because individuals in whom they are absent or defective have mental retardation. Recently, scientists have begun to map and clone some of these genes, and they have set out on the challenging journey to understand their role in brain function.

In our society, persons with Down syndrome (first described by Langdon Down, a British physician, 150 years ago) represent the archetype of mental retardation. This is because affected individuals look remarkably alike and because the syndrome is so common (appearing in about 1 in 1000 births). Yet, no one even had the slightest guess as to the cause of Down syndrome until 1936 when Lionel Penrose, a British geneticist, offered conclusive evidence that the chances of bearing such a child rose dramatically as women reached their late 30s. Over the ensuing 20 years, this finding spawned much conjecture, but few new facts. The big advance came in 1959 when Jerome Lejeune, a French geneticist, found that persons with Down syndrome invariably have an extra chromosome. This extra chromosome, by convention number 21 (the chromosome pairs are numbered in descending order of size, so this is almost the smallest), is present due to an error in the way chromosomes separate during the formation of egg or sperm cells. About 5% of persons with Down syndrome have the disorder for another reason: In one of their parents the extra chromosome is stuck to another chromosome. That parent has the right amount of genetic material, but it resides on only 45 chromosomes. If the fused chromosome gets passed on, the child will have the normal count of 46 but will have 3 chromosomes numbered 21. In a few cases, such persons inherit only part of the extra number 21.

The scores of physical findings associated with Down syndrome (including short stature, small head, thick tongue, flat cheeks, increased risk for congenital heart defects, short fingers, unusual skin folds on the hands, sunny disposition, and low intelligence) all must arise because of the presence during development of the extra doses of the proteins made by the genes on the extra chromosome 21. By studying hundreds of persons with Down syndrome, especially those persons who have only portions of an extra 21, scientists have begun to define a "critical region" within which the genes that compromise intelligence must lie. That is, they have greatly narrowed the search for the genes that cause the mental retardation.

Using new techniques in molecular genetics, in 1996 Desmond Smith, a British physician-scientist working in California, successfully transferred small chunks of human chromosome 21 into young mouse embryos. These animals have grown up and now represent an exciting new resource for refining our understanding of the role each human gene plays in intellectual functioning. Smith and his colleagues have devised ways to test the intelligence and memory of these transgenic mice. They can then correlate their performance with which portion of human chromosome 21 they carry in their genomes. Because they can create chromosomal chunks of different sizes and study their impact on many animals, the scientists have already greatly refined our understanding of the critical region. This will enhance our understanding of how genes in this region work to support the development of normal and even superior intelligence.

Might we someday be able to use genetic engineering to enhance human intelligence? I can imagine a future when humans will be able to assess the capacity for superior intelligence by performing genetic tests on early human embryos and improving that capacity through gene therapy (Chapter 20). One might argue that we have even taken the first steps on that long journey. In the summer of 1999, Dr. Joe Z. Tsien, an assistant professor of biology at Princeton, published an extraordinary paper suggesting that by adding a single gene to early mouse embryos, he had conferred upon the live-born animals the capacity for doing spectacularly well on the equivalent of mouse intelligence tests. It appears that he did this by enhancing their ability to remember. It has long been thought that the capacity to form memories and success in taking tests that are used to measure intelligence are closely related. In mice a complex molecule called the NMDA receptor, which is assembled from several proteins (each coded for by a different gene), is known to have a huge influence on memory formation. Dr. Tsien reasoned that if he added another gene for a protein called NR2B (which is one of the components of the NMDA receptor) to the two already present in the cells of a mouse embryo, the resulting extra protein in the young mouse would have a positive effect on its neural activities, including memory and learning. He was able to create the transgenic mouse embryos which grew into mice that can negotiate a maze much quicker than can their unmanipulated relatives. The basis for their superior performance seems to be to that in the cells of the genetically engineered mice the altered NMDA receptor stays open about 150 thousandths of a second longer than do regular cells. Predictably, the report of a "smarter mouse" provoked an invasion of journalists, all asking when the techniques would be ready for human use.

Even cursory reflection on the immense complexity of the human brain, an organ with billions of cells involved in trillions of interconnections, and operating in an ever-changing pattern, makes the hubris of anyone who would attempt such actions with current knowledge laughable. Still, we may have started on a journey that will someday result in our learning how to genetically enhance intelligence. If so, I am glad it is still a long road, for we are clearly not yet mature enough to be able to use such skills in ways to benefit humanity and its fragile home. Until we mature as a species, it is surely better that the origins of intelligence, musical and artistic talent, athletic prowess, and the thousand other qualities that make us fascinated with (and envy) each other remain elusive.

Dean H. Hamer, Ph.D.

Gay Genes
What's the Evidence?

THE HUNT FOR A GAY GENE

In the summer of 1993, *Twilight of the Golds,* a play by 26-year-old Jonathan Tolins, took San Francisco, where it premiered, by storm. The play depicts a middle-aged Jewish couple painfully struggling to deal with the possibility that their daughter could bear a son with a gene that will make him gay. Having made an uneasy peace with the homosexuality of their only son, the couple is devastated by this threat to their hope for a grandson who will normalize the family. The play, which landed Tolins on "Nightline" and in the pages of *Time,* was prescient. While Tolins was writing it in California in 1992, a molecular biologist in Maryland was trying to see whether the artistic premise was scientifically valid.

During most of his distinguished career, Dean Hamer, the scientist who took on the question, has worked at the National Cancer Institute in Bethesda, Maryland. For almost 20 years he studied the ways in which genes are expressed, what turns them on and what turns them off. Even by scientific standards, his work was esoteric. He spent a decade investigating the regulation of a single gene that makes a protein which binds to toxic heavy metals such as cadmium when they get in cells. By studying what makes the protein abundant when the toxins are present, but barely detectable when they are not, he was exploring the much deeper question of how genes are regulated.

In 1991, at the peak of his scientific career, Hamer found himself at a crossroads. Despite the importance of his research on gene expression, he yearned for new projects involving scientific questions that had a more direct link to the human condition. One day while attending a meeting in Oxford, England, he picked up a copy of Charles Darwin's book, *Descent of Man, and Selection in Relation to Sex.* Hamer remembers that in the same book shop he also purchased a copy of *Not in Our Genes,* written by Richard Lewontin, in which the Harvard biologist argued that scientists do not have the tools to dissect the genetic contribution to complex human behaviors, such as homosexuality. Reading on the long flight home, Hamer was fascinated to find that Darwin had speculated at length about the possibility that human behaviors were highly influenced by hereditary factors (Darwin did not use the word gene, which was coined by the English biologist, Bateson, about 1906). Hamer was, on the other hand, unimpressed with Lewontin's book, which he saw as a political tract rather than a scientific argument. Like everyone else working in the biomedical sciences in the early 1990s, Dr. Hamer felt the pull to respond to the dramatic challenges posed by the HIV virus. His thoughts turned to the possible role of genes in homosexuality.

Over the past half century, a few scientists had studied the genetic contribution to homosexuality in twin, adoption, and family studies. Early research had shown that among identical twin pairs in which one twin was gay, the co-twin was much more likely to be gay than was the case among fraternal twins. Other studies found that the gender orientation of boys adopted away from their biological parents at birth was statistically more likely to match the gender orientation of their biological brothers than their adoptive brothers. The brothers of gay men and the sisters of gay women were also more likely than average to be themselves gay. Such findings cannot completely discount the role of subtle environmental factors, but they certainly suggest that genetic factors are at work.

When Hamer started to look into the topic, he may have been surprised to learn that even in the AIDS era there were only a handful of recent studies trying to assess whether there was a genetic basis for homosexuality. Important among them was the work of J. Michael Bailey, a young professor of psychology at Northwestern University who had studied gay twins. Among 110 gay sets of twins, Bailey found that of 56 pairs of identical twins, 52% of the co-twins were gay, whereas among fraternal twins, only 22% of the co-twins were gay. He later also showed that the brothers of gay/gay twins were more likely to be gay than were the brothers of gay/straight twins, another suggestion that genetic factors were at work. These data were generated by questionnaires; Bailey had asked gay men to report on the sexual orientation of their brothers, a methodological approach that cast doubt on the results. Hamer knew that molecular biology had tools that were much more likely than questionnaires to demonstrate a genetic contribution to homosexuality. He set to work to design a study that would resolve whether there was a "gay gene," and which, if there was one, would also show about where it was located.

In the fall of 1991, The National Institutes of Health approved a proposal and provided $75,000 for Hamer's lab (in which about 10 people worked) to use DNA linkage technology to hunt for a gene that influenced gender orientation. After months of planning and several key conversations with leading gene hunters, the team's final plan was to recruit two groups of gay men: (1) a randomly selected number who would be asked detailed questions about their family history to ascertain whether there were more than the expected number of gays among their male relatives, and (2) a carefully selected set of pairs of gay brothers whose DNA would be studied to see whether they were much more likely than not to share certain stretches of DNA, a finding that could indicate that a particular gene which they had in common had shaped their sexual orientation. Such a finding would be reinforced if later it could be shown that the stretch of DNA they shared was not present in other brothers who were not gay.

While Hamer's scientific team, which included Angela Pattatucci, a feminist interested in whether there was a genetic component to lesbianism, was conducting its research, other scientists were approaching the biology of homosexuality from a vastly different angle. On August 30, 1992, Simon LeVay, a neuroscientist at the Salk Institute in San Diego, published a paper in *Science* claiming that among homosexual men, the shape and size of a region of the brain known as the anterior hypothalamus is indistinguishable from the shape and size in women, and distinctly different from the comparable region in heterosexual men. A year earlier, two other neuroscientists had reported that a different brain region, the suprachiasmatic nucleus, was larger in homosexual than in heterosexual men. Intriguing as they are, however, these anatomical differences, even if they stand up to repeated inquiry, do not really offer any insight into a biological cause for homosexuality. They could, for example, reflect a biological consequence of homosexuality, rather than indicate a cause.

All the anatomical studies used tissue taken at autopsy. LeVay was building on an earlier research that had found two regions of the anterior hypothalamus which were twice as large in heterosexual men as in women. He looked at the same regions in 19 homosexual men who had died of

AIDS. This raises the question of whether the size of the region was influenced by having or being treated for the complications of AIDS, an unlikely, but possible, explanation. Even if the particular area of the brain is comparatively larger in healthy young homosexuals (a question that could only be studied in individuals who died young), that would not explain sexual orientation. The anterior hypothalamus could, for example, be enlarged due to the action of one or several genes, which also in completely unknown ways influence gender orientation. That is, the enlargement of a particular region may merely be a secondary effect. No matter how enticing the anatomy, such studies, entangled in the argument over cause or effect, would never fully resolve the issues. The research could realistically be done on only a few people, it was difficult to find good control groups, the methods used to measure brain region size were controversial, and even impressive differences would not indicate much about fundamental cause. Thus, Hamer's work unfolded as scientific debate was reaching a crescendo. His findings reverberated far outside the scientific world.

By placing advertisements in gay newspapers, Hamer and his team recruited 76 gay men who agreed to enter the study. The scientists compiled detailed pedigrees, carefully questioning each subject about the gender orientation of his relatives. The responses indicated that 13.5% of the gay men's brothers were homosexual, a much higher figure than the background rate of 2–3% male homosexuality currently assumed for the general population. Although this 2–3% figure is lower than the often quoted figure of 10%, it may be more accurate because the higher prevalence figure is based on studies that are old and methodologically flawed. For example, at least one early study which concluded that 10% of men had homosexual encounters in adulthood was based partly on interviews with prisoners. Hamer's assumption of a 2–3% general prevalence is in line with other recent studies.

When Hamer studied the extended families of the gay men they had recruited, he and his team found that there were significantly more gay relatives in the maternal side than in the paternal line. This was especially true for maternal uncles and for cousins who were the sons of maternal aunts. To a geneticist, such a finding sends a clear signal—there may be a gene on the X chromosome that influences sexual orientation.

To track down the hypothetical gene, the team selected 40 pairs of homosexual brothers. They took DNA from each and performed a "linkage analysis," a study that asks whether the presence of a particular condition or trait (in this case gender orientation) is strongly associated with a specific tiny bit of DNA (a marker) in each family member who has that characteristic (and absent in those who do not). Because during the formation of egg and sperm long stretches of DNA are transmitted in a block, the association of a DNA marker with a physical trait suggests that a causative gene is located on the same chromosome as, and in the vicinity of, the DNA marker. Brothers have a 50% chance of having inherited the same X chromosome from their mother. If gay brothers are gay because of a gene on the X chromosome, they should share the chromosomal region on which that gene resides much more than the random expectation of 50%. Molecular biologists have sets of DNA probes for determining which of two possible chromosomes a person inherited and whether a particular stretch thereof is present.

Using 22 sets of probes that act as markers on it, Hamer scanned the X chromosome for evidence of linkage. The team found that of the 40 pairs of gay brothers, 33 shared the same stretch of DNA in a region defined by two of the probes that map near the tip of the long arm of the X chromosome. Statistical analysis indicated that the odds of this occurring by chance were about 10,000 to 1. Put another way, if you decided to flip a nickel 40 times with the goal of getting heads 33 times, you could expect to achieve this result on average only about once in 10,000 series. The results were impressive. Nevertheless, it is was clear that the story must be complicated. For starters, 7 pairs of the gay brothers did not share a common DNA marker in the region. Furthermore, Hamer had also found other families with pedigrees that suggested a role for genetic factors on some other chromosome.

Hamer and his colleagues published their findings in *Science.* The paper appeared at the height of the public debate over whether the Department of Defense should amend its policy forbidding openly gay persons from being on active duty in the armed forces, a coincidence that helped generate interest in the study far beyond the academic community. Like most scientific work, the research raised a lot more questions than it answered. If there is a gay gene, how does the protein for which it codes drive this outcome? How does it differ from its corresponding version in heterosexual men? Does the gene have similar effects in women?

The answer to the last question came quickly. The Hamer team recruited two new groups of families, one group with two gay brothers and another with two lesbian sisters, both of which also had heterosexual siblings. By late 1995, they had found that of 32 new pairs of gay brothers, 22 shared that same region at the tip of the long arm of X, whereas the 36 pairs of lesbian sisters did not disproportionately share the same region. Thus, whatever the putative gene is, it does not appear to affect sexual orientation in women. In this study the evidence for an influential X-linked gene was again present, but it was not nearly as strong as in the first study.

In April of 1999, a third paper regarding the possibility that a gene on the X chromosome predisposes to male homosexuality was published in *Science.* The scientific team, led by Dr. George Rice of the University of Western Ontario, reported that among 48 families with 2 gay brothers they found *no* evidence at all to support the existence of a "gay" gene on the X chromosome. The report set off an immediate scientific debate. Dr. Hamer argued that the failure of Rice and his colleagues to find an association between homosexuality and the long arm of the X chromosome was due to the manner in which they had selected the families for inclusion in the project. Dr. Rice and Dr. Neil Risch, the mathematical geneticist with whom he worked, disagreed, arguing that the selection of the families who ultimately participated (from a larger total) was completely random. At the least, the study from Canada suggests that one or more still unknown genetic factors contribute to male homosexuality, as well as an as-yet-unidentified gene lying among 4,000,000 base pairs of DNA on the long arm of X.

THE SOCIETAL RESPONSE

The report of evidence for a gene that predisposes to male homosexuality generated and continues to generate much interest and debate. It is probably the most controversial of the hundreds of associations between human conditions and DNA markers published thus far. For example, the gay community split sharply over the news. Some saw it as proof that homosexual orientation is due to innate differences and thus normal. They argued that the Hamer paper should put to rest all discussion that homosexuality is deviant behavior that arises out of sordid childhood experiences. A T-shirt slogan, "Xq28—Thanks for the genes, Mom," captures the argument. Some gay men even argued that Hamer's work was important evidence in the battle to secure civil rights protection for gay persons. Others worried that society would label them as having a "genetic disease."

When he began the research, Hamer surely never anticipated that it would lead him to a courtroom in Colorado. On November 3, 1992, Colorado voters passed a state constitutional amendment (Amendment 2) that prohibited the state from enacting laws to give homosexuals a special means to secure their civil rights. Those who opposed the amendment, including gay activists, sued, arguing that it was unconstitutional. They reasoned that evidence that homosexuality was a genetically determined (innate) drive over which a person has no control made it arguably analogous to skin color, the condition that has been the subject of most civil rights laws. If a person cannot control his or her homosexuality and if there is an obvious pattern of discrimination against homosexuals in housing, hiring, and other matters, then, they argued, special laws should be passed to provide redress.

On October 16, 1993, shortly after the publication of the article in *Science,* Hamer appeared in a Denver courtroom as an expert witness for those challenging the amendment. He testified that the research suggested that there was an exceedingly strong likelihood that some men were genetically influenced to be gay. His testimony may have helped the gay cause in Colorado. A few weeks later, in an opinion declaring Amendment 2 unconstitutional, Judge H. Jeffrey Bayless became the first judge in the United States to cite scientific evidence in support of the proposition that some men are innately gay. Today, as the confidence with which Hamer could make such assertions has faded, the court's reliance on such limited scientific evidence seems at best naive.

The argument over whether homosexuality is an inborn or acquired characteristic is hardly new. In early 20th century Germany, hoping that it would defuse discrimination and promote tolerance, homosexuals argued that their condition was innate. Their arguments were of course futile. The Nazis agreed that homosexuality was congenital, but also concluded that it was an incurable defect that must be eliminated from the race.

Some gay men have asked why Hamer's work was being undertaken (with taxpayers' dollars) at all. If we are seeking to build and maintain a society that tolerates gays and lesbians, what reason is there for hunting for a genetic basis for their homosexuality? As one writer put it, why don't the scientists hunt instead for a gene that causes homophobia? The implications of this challenge are obvious, and they recall the menacing issues in *Twilight of the Golds*. If we locate gene variations that predispose some young men to become homosexual, we will be able to test for it. Women who are the sisters of gay men will be able to ask whether they carry a gene that would, if passed on, predispose their sons to being gay. If a woman is found to be a carrier, she will be able, if she so wishes (unless society forbids it), to undergo prenatal diagnosis to determine whether she is carrying a male fetus with the predisposing allele.

Would some women abort a pregnancy because the fetus has a gay gene? Almost certainly. Shortly after Hamer's initial report about a gay gene on the X chromosome became public, a woman called me to ask if she could be tested to see if she "carried" the gene. Two years earlier the woman had cared for a much loved older brother as he was dying of AIDS. After hearing about Hamer's research, she had looked into her family history and learned that she had an uncle who had never openly admitted his homosexuality. Now in her early 30s and about to marry, but still grieving the loss of her brother, the woman told me frankly that if there were a carrier test she would take it, and that if she turned out to be a carrier, she would seek prenatal diagnosis. If she learned that she was carrying a male fetus with the gene, she said she would have an abortion. Why? She said that she could not bear the thought that someday she might have to watch a son die of AIDS as she had a brother.

Genetic Testing Issues

Should a woman have the right to find out whether her fetus carries a gay gene? Should a physician help her to find out? Those were the two questions that Mike Wallace really wanted me to answer when he and his crew from "60 Minutes" arrived in my office a few months after the news about a gay gene broke. When the staff at "60 Minutes" set up the interview, I had been told that it would be a wide-ranging interview about advances in human genetics. So it seemed at first. But after a first hour of "softball" questions, during which Wallace seemed to be charming me (of course not a second of this footage ever made it to television), the questions took on a confrontational tone.

Wallace knew that I supported a woman's right to find out information about her pregnancy and to terminate her pregnancy for whatever reason she wanted, even if it was morally repugnant to me. The first couple of times that he asked me, "Now, doctor, would you help a woman find out if her fetus was going to grow up gay, knowing that she might abort the pregnancy?", I dodged the question. After all, it was not and is not possible to provide such information, and it may never be. Genetics deals with statistical likelihood. I knew that further research in followup to Hamer's would most likely show that there were many men who had the "gay" allele at the tip of the X chromosome, but who were heterosexuals. Based on years of experience with other predisposing genes, the most likely possibility was that being born with this particular genetic variant did not program one to be gay. It merely increased the odds that this outcome would occur. Given that Hamer had sought out families with many gay members, it is almost certain that larger, population-based studies would show that if there is a predisposing allele on the X chromosome, it explains a much lower percentage of male homosexuality than originally seemed to be the case. Other unknown factors must play a major role in determining whether a boy with a predisposing gene grows up to be gay.

Of course, Wallace did not want to hear complex arguments about gene penetrance and the relative contribution of nature and nurture. When I said there was no such test, he made the issue unavoidable. "Doctor, suppose there was a test for a gene that research had showed absolutely that if a boy is born with it he will grow up to be gay. If you were a doctor who did amniocentesis (the procedure by which physicians use a large needle to take a sample of amniotic fluid from pregnant women, fluid which has fetal cells that can be tested) and a woman came to you asking that you arrange for DNA testing for the gay gene, would you help her?"

A moment of truth had arrived. In an instant, many thoughts careened through my brain. I thought about a gay friend. I thought about the gay community. I thought about discrimination. I thought about my views on abortion. I thought about pregnant women. I thought of how my answer might seem, especially after editing, and juxtaposed to the views of others. I realized that Wallace might be setting me up as the bad guy, the doctor who would conspire with a woman with irrational prejudices to provide information that could lead to the abortion of a healthy fetus. I had only two choices. I could dodge the question and then terminate the interview (with cameras rolling) or I could answer. "Yes," I said, and quickly continued. "Each year in this country there are about 1,000,000 abortions. In 99% of the cases, women are ending their pregnancies because they do not want to have a baby, almost certainly a baby that would be healthy. They have a legal right to do that. I think that if they have a legal right to end a pregnancy for whatever reason they want, it has to include reasons that I might find repugnant."

Prenatal testing already permits a woman to find out a few important facts about her fetus, such as whether or not he or she has Down syndrome. In the not-distant future, women will be able to find out a great deal more information about their future children. Does the fetus have a gene that predisposes to Alzheimer disease? To breast cancer? Is he or she likely to be obese? Is he or she likely to suffer from serious depression? The list of potential tests is staggering, but most of them will deal with possibilities rather than with certainties. As bioethicist Eric Juengst has put it, the genetic counselor of the future will be more like a weather forecaster than a soothsayer.

Because its use is so closely linked to abortion, our society is already bitterly split over whether prenatal testing is morally permissible. Who can draw the line that separates prenatal tests for conditions that are medically "serious" (and for which testing is defensible) and those which are not (and for which testing may be ever so much more difficult to defend)? It seems to me that more is lost than gained by making physicians moral gatekeepers of these tests. I think that a woman should be able to ask any question she wants about her fetus. In the few minutes that physicians have to counsel about such tests, it is impossible to discern what motivates the woman, nor should it necessarily matter. If her wishes are driven by erroneous understanding of medical risks, that should be addressed, but to educate, not to reshape behavior. There can be little doubt that someday prenatal tests will be available for genetic traits that predispose to conditions that few of us think should drive a decision to end a pregnancy. One study found that women would be about as interested in testing a fetus for the risk of severe obesity as they would for cystic fibrosis! The right course, I think, is not to deny access to testing, but to make sure that women who seek prenatal tests clearly understand the implications of the results before they undergo testing.

PART 4 PLANTS AND ANIMALS

Genetic Engineering and Nature

Transgenic corn growing next to a traditional crop. The larger, healthier-looking corn has a transgene to protect it from a common predator that is harming the standard corn. (*Courtesy of Pioneer Hi-Bred International, Inc.*)

13

Genetically Modified Organisms
The Next Green Revolution?

COCKTAIL PARTY

At first glance, it was just another cocktail party at the John R. Brown Convention Center in Houston, indistinguishable from the thousand that had preceded it in the large, nondescript room. Two hundred or so mid-level executives, a tiny fraction of the thousands who were attending the 1996 convention of the Biotechnology Industrial Organization (BIO), milled about, trying to balance a plastic wine glass in one hand and a paper plate piled with food in the other. One could barely hear the classical guitar music over the din. But, despite outward appearances, the event held extraordinary meaning. This was, quite possibly, the world's first genetically engineered cocktail party! Virtually all the hors d'oeuvres in the room had been prepared from genetically modified organisms (GMOs)! I resolved to sample each of these wonders of molecular biology.

On one table was a gargantuan bowl of corn chips and a bathtub-size bowl of salsa. The chips were made from corn flour grown from seed into which had been inserted a gene that blocks the action of a herbicide with devastating killing power. Sold under the name Liberty, this poison will destroy virtually any broadleaf plant on the planet unless a gene which produces an enzyme that deactivates it is in the plant's genome. If they sow corn seed with the protective gene, farmers can harvest more bushels per acre at lower cost. Why? A single application of Liberty will kill virtually all the weeds on that acre. The corn chips looked, felt, and tasted like every other corn chip I have ever eaten. The salsa was too spicy.

At another table was a metal serving dish holding a mountain of new potatoes that had been roasted with garlic. These NewLeaf potatoes, a product of Monsanto, have been genetically engineered to resist one of the banes of potato farming, the Colorado potato beetle. The protein produced by the single gene that Monsanto scientists cleverly slipped into the potato genome renders the leaves of the potato plants, which the beetle normally finds delicious, to be extremely distasteful to the insect. The roasted new potatoes tasted great.

By the time I had threaded my way through the crowd to the third food table, the bruschetta, a round Italian bread with tomatoes, red peppers, and olives baked into it, was almost gone. This dish was made with Endless Summer tomatoes, the product of a company called DNAP. The company scientists had developed a method to insert a gene into the tomato genome that slows the action of a key fruit-ripening enzyme. This greatly extends the shelf life of the tomato, making it easier and less expensive to keep ripe tomatoes in the grocery bins. Since the genetic system that controls the way fruits ripen is the same in many species, including pineapples and bananas, the benefits of this extended shelf life will be vast. My last dish was vegetable tempura. The yellow squash that was coated in the delicate crust had not had any foreign genes inserted into it, but the canola oil in which the tempura was cooked had been harvested from plants that had been genetically engineered to have increased yield (more oil per plant) and vigor (more resistant to a virus). I was elbowed away from the table by a herd of sales reps before I had eaten my fill.

TRANSGENIC CROPS

During the last five years, genetic engineering—the controlled transfer of a gene or genes from one organism to the germ plasm of another—has immensely transformed American agriculture. Almost overnight, a sizable percentage of the tens of millions of acres committed to the production of soy beans, corn, and cotton have been planted with seeds that have one or more "foreign" genes incorporated into their genomes. The giant companies that dominate agribusiness (Monsanto, Cargill, DuPont, and Novartis are among the leaders) claim that such genetically engineered seed promises the next green revolution—an extraordinary increase in crop yields that will deliver humanity more food at much lower cost.

First in Europe and now in the United States, however, an increasing number of scientists, environmentalists, bioethicists, consumer groups, and small farmers have forged coalitions to challenge the wisdom of transforming world agriculture so rapidly. Their two most important concerns are that there has not been enough attention to food safety and that some GMO plants may carry genes that if transferred into closely related, but wild species, could create superweeds. In the United States, public awareness, relatively small until 1999, is rising rapidly. Consider, for example, the potential impact of a full-page ad created by a consortium of 60 public interest groups including Greenpeace and Friends of the Earth that appeared in the *New York Times* in October of that year. Under the large point question: "Who Shall Play God in the Twenty-first Century?," the ad asks who gave permission to the biotech industry "to put human genes into pigs, fish, and plants?... to put fish genes into tomatoes? to create plants that can't reproduce? to redesign and clone animals (and, soon, humans) to fit a market function? to take over Nature's work?... When did we approve all this?... Have we lost our sanity?"

Despite such scary ads and what will be an intense political battle, GM foods are here to stay. During the next few years we can expect significantly more premarket testing of GM foods. Almost certainly, there will be new laws and regulations requiring many products made from such crops to be so labeled. But the benefits of this technology far exceed the risks.

What is actually going on? Monsanto, perhaps the world's most influential agricultural conglomerate, has driven much of the change, creating dozens of genetic modifications to important crops, each intended to enhance production. For example, Monsanto's genetic engineers have created a corn called *YieldGard,* which contains genes from a bacterium called *Bacillus thuringiensis* (Bt) that make the plant resistant to attack by the European corn borer beetle. Bt is a soil organism that during the course of evolution has acquired a number of genes which make proteins to kill insects that prey upon the plants which it calls home. During sporulation (reproduction), Bt produces enzymes which kill beetles that eat those plants. The bacteria's enzymes are a sort of time bomb. They are activated by enzymes in the insect's gut. Once activated, they kill the insect by destroying the cells in its digestive tract.

Farmers have been spraying crops with Bt for 40 years, but it was only after the first Bt "insecticide" gene was cloned in 1981 that botanists began to use the genes to build resistance to insects into the plant genome. Today, scientists have an arsenal of more than 100 Bt genes, cloned from a variety of bacterial strains, each of which kills a particular type of insect. By incorporating these genes into plants, the molecular biologists have redefined insecticides. Instead of spraying millions of acres with toxic chemicals that defoliate virtually every unprotected plant they hit, farmers are now using natural, biological insecticides derived from soil organisms that humans have been ingesting with food for the entire 10,000-year history of agriculture. The use of seed genetically engineered to carry a Bt gene should be great news to those most concerned about the health and safety of our ecosystem.

GM seed is being planted at an amazing pace both in the United States and around the globe. In the United States, transgenic tomatoes are available in the grocery bins, and the Department of Agriculture has approved the production of dozens of transgenic plants in addition to soy, corn, and cotton. In China, farmers already tend vast acreages of transgenic cereals. In Japan and southeast Asia, there will soon be millions of acres committed to transgenic rice.

The dramatic change in how we produce key crops depends on two new technologies, both of which emerged in the 1980s: (1) plant regeneration (cloning) and (2) gene transfer. During the 1980s, scientists, who had been working on the problem for decades, perfected methods to remove cells from plant embryos and to turn each into a fully formed plant by culturing the cells in a broth of nutrients that included growth hormones. Thus, the embryo of one desirable plant could become the parent of billions of cloned (genetically identical) offspring.

The development of gene transfer technology allowed plant scientists to intervene in a precise manner to genetically modify individual embryo cells that could then become the parents of countless other plants, each of which would also contain the newly inserted gene. Among the first important advances was the discovery that a bacterium called *Agrobacterium tumefaciens* (At), a ubiquitous bug that has the natural capacity to penetrate the cell wall, was an excellent vehicle with which to transfer a desirable gene from another species. Since the 1970s, microbiologists have been able to use a family of enzymes to cut DNA molecules at precise spots as a chemical tool kit for splicing genes in bacteria. Using the same tools, it was relatively straightforward to transfer a potentially desirable gene that had been isolated from some organism into *A. tumefaciens* (the name comes from the fact that it causes noncancerous tumors on the roots of plants that it invades). The genetically engineered At bug could then be placed into culture with the plant cell of interest and act as a vector to deliver the gene it carried into the plant.

In 1994 Japanese scientists succeeded in using *A. tumefaciens* to create transgenic rice (so called because a gene from a foreign species has been transferred into it). Since then, scientists have succeeded in using a similar approach to create a variety of transgenic plants including corn and cassava (a widely consumed tropical plant with starch-rich tubers that is used to make manioc and tapioca). The At vector system has provided botanists with a way to study the impact of particular genes on the size, color, ripening time, resistance to infection, and nutritional qualities of most agriculturally important crops.

In the last few years, scientists have also found other ways to transfer genes that do not depend on using a bacterium to infect a plant cell culture. One, called electroporulation, is based on the fact that when a cell is exposed to a mild electric current, tiny holes appear in its wall through which DNA molecules can pass. Another attractive method (because the outcome may be better controlled) is to bombard plant cells with microscopic missiles that carry a payload of DNA (a gene of interest). Because in both cases the target is a cell that can be cloned, one does not need a terribly efficient success rate. When scientists find a few cells that have integrated the transgenes into their genomes, they can use them to propagate an infinite number of copies.

Botanists began creating transgenic plants with Bt genes in the mid-1980s, and by 1990 they had shown that the proteins coded by the genes protect the tobacco, tomato, and potato plants from certain pests. Since then, they have made remarkable progress by redesigning the Bt genes to produce a much larger amount of insecticide protein. Francis Rajamohan, a scientist at Ohio State University, and his colleagues were among the pioneers trying to genetically engineer Bt to become a super-killer of insects. They study insects with receptors in the cells in their gut that bind Bt toxins. They are trying to re-engineer the Bt genes to make the proteins bind even more efficiently to those receptors, thus causing more devastating injuries to the insects. Their goal is to make new Bt genes that will kill other pests such as gypsy moth larvae, mosquitoes, cabbage loopers, and cotton worms.

The development of Bt corn by no means signals the end of the havoc wrought by the corn borer beetle and other pests. In the intricate dance of evolution, struggling under the harsh conditions created by Bt corn, the beetle is likely to benefit from the appearance and spread (through reproductive success) of mutations that counter Bt. Other pests have done so.

The boll weevil which attacks cotton, and the northern and western corn root worms which devastate corn, are so far invulnerable to Bt proteins.

To feed the world efficiently, farmers will need plant scientists to come up with other weapons against pests. Attacks by root worms and cutworms are so devastating that American farmers spend $1,000,000,000 a year on chemical insecticides to repel them. Scientists are busily testing other soil bacteria to find those which have natural insecticides, and which might lead to a transgenic corn plant that is so resistant to the root worms that farmers will not need to spray millions of acres with chemical insecticides.

The boll weevil, which deposits its eggs inside the cotton flower buds, where they cannot be reached effectively by chemical sprays, does so much damage that it strongly influences the geographic distribution of U.S. cotton production. Botanists have recently discovered that a strain of the ubiquitous bacteria *Streptomyces* contains a protein known as a cholesterol oxidase that is acutely toxic to the larvae of boll weevils. This may lead to a product that will be among the second generation of transgenic insecticides.

THE DEBATE OVER GMOS

A now defunct company called Calgene played a key early role in bringing genetically engineered crops to the American consumer. Its FlavrSavr tomato is a transgenic fruit with foreign genes that allow it to stay on the vine longer and, thus, be tastier than conventional tomatoes which are picked green and, as they are moved across America in box cars, gassed with ethylene oxide, which produces the natural red color by unnatural means. In 1989 Calgene embarked on what became a five-year dialog with the U.S. Department of Agriculture (USDA) and the Food and Drug Administration (FDA), culminating in April of 1994 when an advisory panel decided that the transgenic tomato was not a new substance, that it did not have unusual levels of any toxic substances, and that it did not pose risks of food allergies. Despite their confidence in the FlavrSavr tomato's safety, the panel was concerned that so few people appreciated the magnitude of their decision. As Joan Gusow, a Columbia University professor put it, "We are changing the relationship between humans and nature on a scale of the industrial revolution." She is right.

Within eight months after Calgene got the go-ahead, five other companies had gained approval for seven additional products, including the Monsanto potato that I ate in Houston. Another important approval that came in the mid-90s was the decision by the USDA to permit marketing of a genetically engineered strain of yellow squash called ZW-20. The transgenic squash is resistant to two plant viruses that have caused major economic losses for squash farmers in Georgia and Florida. During the late 90s there was a steadily increasing pace of government approval that was largely due to the fact that the transgenic plants are all created in roughly the same way. New technical and safety issues did not emerge with each application.

Despite the enthusiasm of American agribusiness, transgenic crops provoked concern almost immediately upon their arrival in Europe. The reaction in Europe was in part due to the traditionally much stronger influence of environmental issues on political discourse and in part due to the deeply felt unease about "mad cow disease" in Britain at that time. During the mid-90s the environmentalist Green Party, led by Euro-Parliamentarian, Hiltrud Breyer, repeatedly raised questions about the safety of GMOs and argued that such products are unnatural.

In 1996 a simmering debate boiled over when the European Commission granted Monsanto permission to sell 200,000 tons of genetically engineered soybeans to food manufacturers in Europe. The EC made its decision after advisors concluded that there was no safety issue significant enough to require that the final products (ranging from baby food to ice cream) using the transgenic soy beans be labeled. Nevertheless, in October, EuroCommerce, the industry association for wholesalers and retailers, insisted that member companies must label products to inform consumers if GMOs were present.

In November, 1996, a shipload of genetically engineered soybeans arrived in Hamburg, but the dockworkers refused to unload them, and several major European supermarket chains announced that they would not carry products made with the soybeans, labeled or not. These events were widely reported and had much influence on the public. During 1997 and 1998, the importation of other transgenic products such as genetically engineered corn produced by Ciba-Geigy declined dramatically. By 1998, despite the fact that there had never been an incident in which an individual was harmed by consuming transgenic food, the battle was over. In Britain, GMOs became known as "Frankenstein foods," and many major food retailers publicly promised not to knowingly carry products made with them. In Europe, at least for now, society has voted overwhelmingly against genetically modified food.

During 1999 controversy over GMOs erupted in the United States. The major arguments are (1) that GMOs are unnatural, (2) that moving some genes into food plants creates an unknown risk of food allergies, (3) that transgenic foods may violate certain food purity laws that are part of the Orthodox Jewish and Islamic religions, (4) that transgenic crops may pose harm to particular species, (5) that transgenic crops may upset existing ecological balances, and (6) that GMOs will be used by agribusiness to the economic disadvantage of the third world.

GMOs are unnatural—in the sense that they exist only because humans are capable of creating them. But this is also the case for virtually every other agricultural product used by humans. All our crops and livestock have arisen from an intensive campaign of forced breeding that has been under way for hundreds of generations. As for concerns about allergens, there is a small risk of creating transgenic organisms that will provoke allergic reactions. For example, if genes from peanuts were moved into corn, corn might become a dangerous food for those people with a severe peanut allergy. I know of only one occasion in which it was proposed to move a nut gene into a cereal, and that was abandoned when such concerns were raised. The number of transgenic food products that have been engineered in a way that eating them would violate a religious law is vanishingly small. Fortunately, both the danger of allergic reactions and the risk of causing people to violate their religious obligations can be solved by labeling, a topic to which we will return.

In the United States, concern for GMOs burgeoned in May of 1999 when researchers at Cornell University reported that in laboratory experiments they had found evidence that the pollen from Bt corn killed monarch butterflies, one of America's favorite insects. Monarchs dine exclusively on milkweed, a plant that is found throughout the midwest, frequently adjacent to corn fields. The Cornell scientists surmised that Bt corn pollen must frequently be blown onto milkweed. In the laboratory, about half the monarchs that dined on milkweed dusted with Bt pollen died, while none of the monarchs that ate milkweed without Bt pollen died. Given that more than 20,000,000 acres of American soil is planted with Bt corn, the experiment suggested that the transgenic corn threatened the butterflies.

The study, which was reported on the front page of the *New York Times,* brought the issue of how little we have studied the ecological impact of GMOs squarely before the public. Dr. Margaret Mellon, director of the agriculture and biotechnology program at the Union of Concerned Scientists, opined, "This should serve as a warning that there are more unpleasant surprises ahead." Overlooked in public discussions about the threat that Bt corn may pose to the monarch butterfly is the fact that unusually severe storms in the Mexican mountains where most of the species spends the winter have occasionally decimated the population, but it repeatedly bounces back. Deforestation of that environment is a far more real threat to the monarch than is Bt corn. Without a protective forest canopy, the monarchs cannot survive severe winters. During 1999 botanists and molecular biologists hotly debated the implications of the butterfly research, with most asserting that Bt corn poses no significant threat to monarchs in the field. Nevertheless, the debate focused attention on the limited review that currently precedes government approval of transgenic crops. Also lost in the rush to criticize GMOs was the fact that by planting GM seed, farmers have sharply reduced their use of chemical pesticides.

The major ecological worry about transgenic crops is whether there is a risk that transgenes that confer resistance to herbicides could (via windblown pollen) cross into closely related species in the wild and create new strains of weeds that spread rapidly, creating a menace to the environment as, for example, occurred with the importation of kudzu, which covers much of the South, and purple loosestrife, a beautiful aquatic weed which is almost impossible to kill and which is choking the ponds of New England. Unfortunately, researchers cannot design experiments that assess such risks, so the environmental concerns cannot be laid to rest. We will have to work with indirect assessment methods and make the most of experience.

As public awareness of GMOs rises and opponents stoke the debate, the public will want to be more informed about the methods used to create the food which makes its way onto supermarket shelves. They will claim a right to know and demand labeling of GMOs. Of course, labeling adds to the price of food in the marketplace, which raises the question of whether the price of disclosure may deprive some people of having the option of consuming these products.

It is likely that over the next few years the USDA will require that there be more extensive field tests of transgenic plants aimed at assessing ecological concerns, and it is possible that foods derived from at least some transgenic crops will be permitted to market only if they carry a label informing the public of their origin. Given the extensive manipulations that humans have performed on our food chain for centuries, I doubt that GMOs carry any more risk than the foods that preceded them. On balance, given the significant reduction in the use of chemical pesticides that they permit, I think it likely that they will be an ecological blessing. Given the immense benefits they promise and the paucity of data suggesting health or environmental risks, GMOs are here to stay. Nevertheless, the public anxiety over GMOs offers an important lesson about the need for advance dialogue with society when technocrats begin to contemplate tampering with something as fundamental as the food supply.

PLANT FACTORIES

The emergence of transgenic plant technology raises the fascinating possibility that plants could be designed as factories to produce at low cost otherwise extremely expensive medicines and other crucial products. The first signs of future success are emerging. In 1996 CropTech, a tiny biotech company in Virginia, announced that, using *Agrobacterium tumefaciens* as a vector, it had successfully spliced a gene that codes for an enzyme called human glucocerebrosidase into the genome of tobacco plants. This enzyme is the only effective treatment for Gaucher disease, a rare genetic disorder that severely affects about 30,000 persons in the world. Some children who are born without the ability to make proper glucocerebrosidase grow poorly, develop big spleens, and have severe anemia. Genzyme, one of the largest biotech companies, currently derives much more than half of its revenues from sales of Ceredase, a genetically engineered form of the enzyme. The cost to treat a single patient with this drug ranges from $100,000 to $300,000 a year, making it a contender as the world's most expensive medicine.

During the late 1980s and early 1990s, Genzyme manufactured Ceredase by extracting it from the cells of thousands of human placentas collected for it by a company in Belgium. It now has a better way to make the drug. Genzyme scientists transferred the human glucocerebrosidase gene into bacteria, which are grown in vast numbers in gigantic fermentation tanks. The new product, Cerezyme, is extracted from the gleaming steel tanks in a "factory" in Allston, Massachusetts. The company no longer depends on a large supply of human placentas, but the huge costs in product development mean that it will not lower the price of the medicine any time soon.

Carol Cramer, CropTech's vice president for research, claims that a single genetically engineered tobacco plant produces about one milligram of enzyme per gram of fresh leaf. This means that a single plant could produce enough enzyme to provide the medicines to treat one person for one year. Theoretically, the reduction in production costs from using plants as factories could be dramatic and would greatly benefit persons with Gaucher disease. We are years away from knowing whether CropTech's vision will be realized.

The genes of the much maligned tobacco plant are among the most studied in the world. Because we know it so well, it may emerge as the key plant in which to design protein factories. In 1997 French scientists announced that they had successfully completed a three-year project to transfer the human gene that codes for hemoglobin, the protein that carries oxygen through the body and makes blood red, into the tobacco genome and to extract tiny amounts of human hemoglobin from its leaves. The main impetus to produce hemoglobin from plants is to produce a blood substitute that carries no risk of threatening patients with lethal human viruses such as happened with HIV in the 1980s. For reasons relating to the cost of production and yield per plant, tobacco does not appear to be the best choice for hemoglobin, but it provided a wonderful test system. The French scientists are now trying to do this work in corn.

Even more exciting is the prospect that within a decade we may be able to give children their immunizations simply by having them eat specially engineered, uncooked vegetables. More than a decade ago, Hilary Koprowski, a biologist at Thomas Jefferson University in Philadelphia, developed a genetically engineered rabies vaccine in edible plants. In 1997, working with immunologists, he created a double vaccine by infecting plants with viruses with genes from both the rabies and HIV-1 virus. He fused genes from the two lethal viruses with the gene for the coat protein of the benign alfalfa mosaic virus, and then put the resulting "construct" under the genetic control of the tobacco mosaic virus. Tobacco plants infected with the modified virus produced large quantities of viral-like particles that comprised the outer coat of the alfalfa virus which enclosed the proteins from the dangerous viruses. When these particles were purified and injected into mice, they made them immune to rabies.

Currently, there is great interest in turning the banana into a vector for vaccines. Given the dismal poverty that engulfs much of Africa, it will be years before there will be a public health infrastructure to deliver traditional vaccines to rural areas, even allowing for the massive ($700 million) gift that the Gates Foundation made in 1999 to overcome this problem. If one could grow bananas and other plants that had been genetically engineered to make vaccines, one could reach most of the population within a few years.

Genetic engineering may someday solve the energy problem by unlocking the virtually inexhaustible energy supplies stored in growing plants. The world runs on oil, vast underground lakes of hydrocarbons created over millions of years as the earth's crust pulverized decaying plants. Rather than exhausting the planet's oil supplies (which at current rates probably will last a couple of centuries), it is not unreasonable to hope that we may someday derive fuel directly from the hydrocarbons that make up much of the biomass in plants. Genetically engineered bugs may be the answer, permitting us to grow our fuel which they will ferment for us.

Xylose, a five-carbon sugar, constitutes nearly 30% of the world's biomass. If it could be inexpensively converted to ethanol, it could greatly ease demand for fuel. In Brazil, for example, a substantial fraction of the trucks and cars run on alcohol fermented from sugar cane. The key to success here is a bug called *Zymomonas mobilis,* which is best known because it ferments the sugary sap in the century plant (*A. tequilana*) to make tequila. But *Z. mobilis,* which is really good at breaking down six-carbon sugars, cannot ferment five-carbon sugars. In 1994 Stephen Picataggio of the Department of Energy's National Renewable Energy Laboratory (NREL) in Golden, Colorado, used gene transfer techniques to create a strain of *Z. mobilis* that can break down xylose. He isolated (cloned) four genes from *E. coli,* a bug found in the human gut, which together produce proteins that catalyze the chemical steps to turn xylose into ethanol. He then inserted them into the *Z. mobilis* genome. The new strain could be used to ferment anything from corn cobs to municipal garbage into ethanol. The NREL is investigating whether a tough weed called switchgrass that covers hundreds of thousands of acres on the American prairie might be genetically engineered to produce massive amounts of ethanol at acceptable cost.

WINE SLEUTHS

Genetic engineering may soon be used to create even better grape cultivars than man has developed through centuries of trial and error. In 1997 biologists answered one important question: What is the origin of the prized Cabernet Sauvignon grape, source of a wine renown for its pigments, tannins, and aroma? Sometime in the 18th century, this dark grape appeared in southeastern France. Within a few decades, viticulturists had adapted it to the climate around Bordeaux to produce an immensely popular wine. As its popularity grew, so did speculation about its origin. Some guessed that it was the long-lost Biturica cultivar praised in Roman times, others guessed it originated in Spain, still others along the Adriatic coast.

When Carole Meredith, an ecologist at the University of California at Davis, was asked by a group of California wine growers to use molecular genetics to verify the identity of key stock plants (a science known as ampelography), she had no idea she would solve one of oenology's favorite puzzles. To develop a reference base, she and her colleague, John Bowers, performed DNA fingerprinting on about 50 cultivars by studying variation in microsatellite DNA (short, highly repetitive stretches of DNA that vary hugely in size among varieties), including all of the common, and most of the obscure, varietals used worldwide in the industry. Analysis of the various stocks showed that all the microsatellites in Cabernet Sauvignon derived from the Cabernet franc or the Sauvignon blanc plants, both native to Bordeaux! All along the name had correctly depicted the parentage, but no one had been sure. Cabernet Sauvignon probably arose accidentally, for even 200 years ago grape growers would probably not have attempted a cross of those particular strains because experience taught them that it would not be expected to retain its all-important disease resistance.

More recently, Dr. Meredith and her colleagues have used DNA analysis to show that all 16 types of wine grapes grown in northeastern France (including Chardonnay, Gamay noir, Aligote, and Melon) derive from a single pair of parents, the Pinot and Gouais blanc, both of which have been cultivated since the Middle Ages! The attribution of parentage to the Pinot came as no surprise, but the historical importance of Gouais blanc will shock the grape growers. Gouais is widely thought to be an inferior grape and is today rarely cultivated in France. We may soon be drinking genetically engineered wine. The creation of transgenic grapes may prove to be the most efficient way to create strains with better disease resistance. It is, however, much more difficult for me to imagine the use of transgenes to enhance taste, which is as much a product of the subtilties of soil and sunlight as it is of genes.

NUTRITIONAL GENOMICS

What will be the most important development in plant genetics in the next 10 to 20 years? I would place my bet on nutritional genomics, an infant science that is already commanding substantial resources of giant companies like Du Pont. The vision of nutritional genomics is that scientists will use molecular biology to define those chemicals in foods that really confer health benefits. Next they will engineer plants to contain more of the really helpful compounds and less of those that may be harmful. Finally, they will develop a composite picture of how humans vary in their ability to absorb and metabolize those compounds and develop foods to match the varieties of human need. Indeed, one can imagine a day in which foods are chosen for consumption based on a genetically influenced wellness strategy. In its most advanced form, nutritional genomics will reshape preventive medicine, keying it to the construction of individual diets that will sharply reduce or delay the onset of disease.

This a grand vision, and even the most optimistic nutritional scientist would agree that it will take decades to realize if fully. But the impact that nutritional genomics will ultimately have is so vast that it is not yet possible to perceive its limits. For starters, molecular biologists and nutritionists will be able to provide a much more accurate picture of the benefits of the food we eat. Indeed, they will in time redefine how we think of food. Breakfast cereals may, for example, be manufactured in a manner that counters one or more genetic predispositions to heart disease. Fruits may be genetically engineered to attenuate a known genetic risk for colon cancer. There may come a day when food manufacturers market directly to consumers based on wellness claims that tie the genetic status of the consumer to the genetic profile of constituents in the food. One biotech company is already developing an assay to screen the impact of key food ingredients on the body's inflammatory response. If foods could be designed to modulate such systems, they could open up new approaches to managing chronic disorders like arthritis and treating less common disorders like lupus. Designer foods will become a reality.

Nutritional genomics could come to occupy a middle ground between the pharmaceutical industry, which will be the source of many new medicines based on genomic research, and the alternative medicine movement that is robust today. Millions of Americans today consume St. John's wort, echinacea, ginko, and hundreds of other relatively unstudied substances to ward off diseases or retain health. There is, no doubt, real medicinal value in some compounds contained in these substances. The tools of nutritional genomics will allow us to isolate them and understand their benefits and their modes of action. Nutritional genomics may provide a much-needed bridge between traditional medicine and holistic health care.

A transgenic goat which carries a gene that produces an important human protein that can be harvested from her milk. (Courtesy of Genzyme Transgenics Corporation.)

14

Transgenic Animals
New Foods and New Factories

A Fish Story

In 1994 Robert Devlin, working at the Canadian Department of Fisheries and Oceans, inserted a fish growth hormone gene into the eggs of the Pacific coho salmon. After fertilization, about 6% of the resulting embryos took up the gene and developed into fish with double the natural supply of growth hormone. The results were astounding. The transgenic fish grew faster and bigger than their natural siblings. A few were truly giants, reaching a weight 30 times greater than the average of coho taken from the sea! Unlike people, fish grow throughout their life span, so their actual weight limits could be even greater. The transgenic fish also matured faster, which is crucially important to efficient food production. When interviewed after their research was published in the prestigious journal *Nature,* the scientists guardedly predicted a 20-year lead time to commercialization.

If Elliot Entis and his colleagues have their way, that prediction will turn out to be terribly wrong. They are fish farmers, not fishermen, fish farmers. And they have a fish story that trumps any of the yarns one might hear on the docks of Key West. The difference is that they have the pictures to prove it. Their company, A/F Protein, Inc., based in Waltham, Massachusetts, is committed to using transgenic techniques to change the way that the Atlantic salmon (as well as several other species, including trout), one of the most popular and expensive fish, reach your dinner table. The goal at A/F Protein is to grow much bigger salmon, much more quickly, in a much cleaner environment, and at a much lower cost. Since its scientists developed a transgenic approach to creating bigger salmon in the mid-1990s, the company has made impressive progress. At the end of their first year of life, its transgenic salmon are four to six times heavier than are one-year-old Atlantic salmon netted in the ocean or nontransgenic salmon grown under identical conditions. Laid side by side, the cultivated fish towers over the others, and virtually all the weight difference is due to muscle.

How were the transgenic fish created? The scientists at A/F Protein achieved these astounding results by inserting an additional copy of the salmon growth hormone gene into early salmon embryos. After cloning that gene, they placed it under the control of a promoter sequence from another species that in normal circumstances regulates the production of one of several proteins called antifreeze proteins (because they protect the northern ocean fishes from death due to near-freezing water temperatures). A promoter is a short stretch of DNA, usually located just before the structural gene, that controls its timing and level of activity. Proteins are made when a regulatory protein binds to the promoter and tells it to turn the relevant gene on. The transgenic Atlantic salmon are not huge merely because they have a new promoter sequence. The real change is that the artificially constructed gene is steadily expressed in the fish liver, not just occasionally expressed in the pituitary. Yet, blood levels of the hormone in the transgenic fish are not abnormally high, probably due to the speed with which the hormone is taken up by cells.

After they souped up the salmon GH gene by giving it a very active promoter, the scientists injected the construct (as it is called) into early salmon embryos. Some cells in some of the embryos took up the gene and incorporated it into their genomes, including cells that would eventually create their reproductive systems. When a pair of the genetically engineered fish born after this transgenic manipulation mate, they pass on two copies of the more powerful growth hormone gene to their offspring. A single fish can have thousands of offspring. In nature most do not survive, but in a fish farm their odds are much improved.

The effort to produce and market transgenic salmon and trout is at the leading edge of the "Blue Revolution," a movement to use sea creatures to feed the world. The United Nations has estimated that in the long run, the only way to sustain a world population of 10 billion people is to vastly increase (at least 7-fold the current level) the food yield from the oceans in a manner that does not deplete them. Land-based aquaculture may hold part of the answer, with the special dividend that it does not want to redefine the oceans as vast watery farms.

How will consumers react to transgenic fish? That depends on many factors, including their perception of its safety, the look and taste of the fish, and its cost. There is no rational reason to worry about whether it is safe to eat. The transgenic salmon has no foreign genes and produces no altered proteins. It does make more growth hormone than its smaller cousins, but that protein is quickly metabolized. On repeated assays, the flesh, which looks and feels the same as ocean-grown salmon, has normal levels of growth hormone in it. How does it taste? A/F Protein recently hosted a dinner at The Stone House, one of Canada's premier restaurants, that featured its transgenic salmon and trout. Several of Canada's most highly regarded chefs sat down to dinner with government officials and tried a variety of fish-based dishes. The verdict? The chefs declared AquAdvantage trout and salmon to be absolutely delicious and indistinguishable from the finest ocean-caught fish!

Such testimonials are not likely to satisfy the "Pure Food Campaign," a coalition of advocacy groups directed by perennial biotechnology critic, Jeremy Rifkind. Like similar groups in Europe, it wants any food product that has in any way been genetically altered to be labeled so that the consumer can factor that into his or her decision to purchase. The management of A/F Protein, Inc. completely agrees. Elliot Entis plans to label his products whether he is required to or not because he is so confident that consumers will find them much more desirable than the more expensive ocean-caught fish. He hopes to satisfy both U.S. and Canada regulatory requirements in 2000 and have the fish in stores by 2001. Price-weary consumers may then see a dramatic drop in the cost of salmon, possibly to below that of beef.

Once they learn more about transgenic fish, at least some staunch environmentalists should be converted. These fish will be cultivated in land-based breeding pens, which are much less dangerous to shoreline environments than typical offshore fish-breeding facilities, which generate so much harmful waste effluent (largely from having too much fish feces in too small a space) that they can seriously harm the microflora of the surrounding shallow ocean waters. Furthermore, there should be less need for offshore facilities, creating less demand to use aesthetically attractive shorelines for commercial purposes.

What is the major objection to the AquAdvantage salmon and the dozen other aquatic species that A/F Protein is genetically engineering to be bigger and cheaper? Skeptics point out that in agriculture, which has progressed slowly over thousands of years, humans have painstakingly accumulated much knowledge about plant breeding. But in the thousands of years that humans have fished, they have never bred genetic changes into any organism. They fear that transgenic salmon, trout, shrimp, and other species will inevitably escape into the larger world, and that such an event could lead to an ecological disaster. For example, Rebecca Goldberg of the Environmental Defense Fund worries that the AquAdvantage salmon could outcompete its nontransgenic cousin and in a few generations take his place in the ecosystem. It is impossible to completely allay such concerns.

Despite their potential benefits to consumers and the environment, a great political battle over the approval of products like the transgenic salmon may be looming. The world-wide scare over "mad cow disease" in the United Kingdom, an animal brain disease that has been linked to fewer than 20 human deaths, less than the number of people killed in auto accidents on a typical weekend in the United States, has generated fears that are hard to calm, fears that have been exacerbated by some journalists. Entis recounts that he foolishly spent an hour educating a British tabloid journalist who was grilling him about safety, only to find that the story headline read, "Gene company denies link with mad trout disease," a disease that does not exist.

Ultimately, the future uses of transgenic fish will be decided politically. In Canada, where both the government and many fisherman are in favor of marketing transgenic fish, the outlook for A/F Protein products is good. In the U.S., and even more so in Europe, governments are more cautious and fishermen are not yet pressing for the product. In the U.S. the major political battleground is currently in Maine. Governor Angus King is boldly defending the state's burgeoning aquaculture industry despite criticism from environmentalists who argue that artificially created salmon that escape into the wild and breed with the few remaining "natural" salmon that spawn in Maine rivers are violating their genetic integrity. Nevertheless, the scientists at A/F Protein are confident. Arnie Sutterlin, who supervises the A/F Protein hatchery in Canada, is fond of reminding folks that virtually no consumer knows that the chicken he eats is a highly manipulated descendant of a jungle fowl in Asia. He is convinced that if the fish is safe, tastes good, and costs less, the world will beat a path to their door.

Hormones as Drugs

The ten-year battle over injecting genetically engineered bovine growth hormone into cows to increase milk production shows how difficult it can be to allay consumer concerns about unfamiliar changes to staple products. Dairy farmers have been striving (with great success) to increase milk production in herds for several hundred years. Long before Darwin offered the intellectual explanation for why what they did worked, farmers routinely selected the cows that were the best milk producers to mother their next generation of calves. A major advance occurred in the 1950s when dairy farmers began to use artificial insemination with semen from prize bulls, but even those production gains pale before the gains with bovine growth hormone.

We have been using hormones as drugs since the discovery of insulin in 1927, but for decades our methods for purifying them were crude and the cost of using them was high. The rise of genetic engineering in the 1970s made it possible to gain access to virtually any human or animal protein in theretofore undreamed-of quantities. The field grew out of the discovery in the late 1960s that many species of bacteria make enzymes that help defend themselves against other bacteria which attack them by chemically cutting alien DNA. The defender's enzyme will cut DNA at a particular recognition site (a short sequence anywhere from four to eight base pairs long) that might appear many times in other organisms, but is not part of its own genome. The discovery of these restriction endonucleases is one of the most critical developments in modern biology. As scientists have discovered more and more of them, they have built a huge catalog of specialized DNA cutting tools. By using those tools in combination, they can cut DNA up in different ways to get sequences of different lengths. This ultimately let them develop efficient methods to find and clip out genes they wanted to study and to mass produce them.

The dairy industry quickly guessed that if cows were given higher doses of their own growth hormone they might produce much more milk. Because growth hormone genes from several species had already been isolated, it was straightforward to clone bovine growth hormone, mass produce it in bacteria, and study its effects in cows. The hard part was satisfying the Food and Drug Administration that it was safe to use, a process that took ten years. The FDA approved Posilac, the brand name for genetically engineered bovine somatotropin (BST), which is another name for growth hormone, on November 5, 1993. Approval meant that a team of neutral expert reviewers had concluded that use of BST to increase milk production would be safe to the cows, to the humans who would consume the milk, and to the environment. Within a year, more than 1,000,000 of the nation's 10,000,000 cows were being injected with BST. By 1997 the number had climbed above 2,000,000. Although there is some debate over exactly how much injection of BST increases milk production per cow, even the most conservative estimates exceed 15%.

Even before Posilac came to market, activists founded the aforementioned Pure Food Campaign to fight it. The initial focus on safety to humans did not generate much public response. BST cannot even be detected in the milk itself, and even if there are minuscule levels, the protein would be almost instantly destroyed in the human gut. Like insulin, it could only be active if injected directly into blood. The much respected former Surgeon General, C. Everret Koop, forcefully asserted that the milk was safe. The impact on cows was not quite so benign. Some animal welfare activists were incensed that cows getting BST were more likely to develop mastitis, an infection of the udders. Mastitis is a common problem in dairy cows, a consequence of the century-long effort to maximize milk output. BST may have contributed in a small way to a longstanding problem, but it did not greatly exacerbate the problem.

The concern that generated the biggest public response was that BST was being used by agribusiness, which could afford it, to put more economic pressure on small dairy farmers who could not necessarily afford the drug. The famous Vermont-based ice cream maker, Ben & Jerry's, even labeled their ice cream containers with the slogan "Save family farms—No BGH." The greatest measure of the success of the consumer activists was in August, 1993, when Congress temporarily banned the sale of the hormone, but that ban was lifted after Monsanto secured FDA approval. Ultimately, the biggest benefit of widespread use of BST may be environmental. If it permits farmers to meet production goals with fewer cows, the nation's dairy herds will release less methane gas into the atmosphere and their manure will pollute fewer streams.

The early use of genetically engineered hormones to treat humans did not trigger nearly the public outcry that the use of BST in cows evoked. This was probably because the hormones were being offered to children for whom no other therapies were available. During the 1960s and 1970s several thousand children with primary growth hormone deficiency who desperately needed human growth hormone if they were to avoid going through life with extreme short stature (a final adult height not much more than that of a 4-year-old) depended on human cadavers for their medicine. The growth hormone, which was always in short supply, was extracted from pituitaries, purified, and injected. Use of human growth hormone from cadavers was banned by the FDA in 1985 when several patients developed a fatal brain disorder (Creutzfeldt-Jakob disease) from a virus that had been transmitted in the extracted pituitary glands.

Serendipitously, in October of that same year, Genentech, among the first and most famous biotechnology companies, received approval from the FDA to market a genetically engineered form of human growth hormone. Cadaver pituitary glands were no longer needed. The new drug, named Protropin, only the second genetically engineered drug to be approved, was manufactured by cloning the gene for human growth hormone and inserting it into a weakened strain of *E. coli* bacteria. Genentech fermented trillions of these bacteria in huge tanks and then purified the hormone (which is chemically identical to the naturally occurring form) from the broth to make a compound that was safe to inject into children. A few children do suffer from side effects, including diabetes and reduced thyroid function, because their bodies develop an immune response to what they sense to be a foreign protein. Protropin and virtually identical versions made by several competitors are now in wide use to treat severe short stature.

Over the last decade, as more people have heard about human growth hormone, many parents have requested it for children who do not have a growth hormone deficiency, but who appear destined to be among the shortest of adults. A growing number of studies have indicated that the regular injection of hGH in some of these children will modestly increase their final adult height. Requests to use the drug to treat routine short stature raises several ethical issues that may be considered surrogates for a similar debate over the scope of gene therapy in the future. At what height is someone so short that there is good reason to treat him or her? Who decides that question? Should anyone who wants to use hGH as a growth enhancer and who is willing to pay the hefty fee for it have the right to use it? Advances in medicine, especially in gene therapy, will blur the line between interventions to treat an illness and those to enhance the capabilities of a healthy person.

ANIMAL FACTORIES

Compared to their concern over consuming the meat of transgenic animals or the milk of cows that have been given genetically engineered drugs, the public seems much more favorably disposed to using transgenic animals to produce precious medicines, says Jim Geraghty, who should know. Until recently, Geraghty was president of Genzyme Transgenics, a company that is attempting to revolutionize the pharmaceutical industry by using transgenic goats as living factories. No such drug is yet on the market, but soon could be.

Genzyme Transgenics operates a factory of the future: Instead of computer-programmed robots, its tools are several hundred purebred goats imported from New Zealand. Since 1990 Genzyme scientists at a farm in central Massachusetts have been creating transgenic goats that carry in their genomes one of a growing number of human genes. They use these animals as the founders for all their transgenic lines, a process that takes about 18 months. In each case, the scientists use the same basic approach. They start with a cloned β-casein gene, which produces the most abundant protein in goat's milk, with which they combine the human transgene of interest. They then use extremely fine glass needles to insert this DNA construct into early goat embryos. After microinjecting the gene of interest into the embryos, the scientists transfer them into the wombs of surrogate mothers for a 5-month gestation. After the surrogate mothers give birth, it takes another 8 months for the kids to become mature enough to themselves become pregnant. After the transgenic goat gives birth, it produces milk that expresses the foreign protein. Eighteen months is about the amount of time that companies devote to building brick and mortar fermentation facilities to produce rare proteins. But using a few dozen goats as one's factory is potentially much cheaper. Once the dedicated transgenic goat herd is created, it takes only 30 goats a year to make 100 kilograms of the protein, about as much as a huge factory might yield.

The first protein that Genzyme Transgenics has manufactured in goat factories is antithrombin-III (AT-III), a blood-thinning protein that is of great potential value in preventing blood clots which can cause both heart

attacks and strokes. By February of 1994, the molecular biologists had succeeded in developing transgenic goats from which they were harvesting
seven grams of antithrombin-III from every liter of goat's milk. The company is now using the same strategy in a dozen or more collaborations
with pharmaceutical companies, including the manufacture of a bone
growth factor, a monoclonal antibody targeted against colon cancer cells,
alkaline phosphatase which is a commonly used test reagent, and several
established protein anticancer agents. Genzyme Transgenics thinks it can
produce rare proteins at 1% of the cost of traditional cell culture facilities.

Efforts to turn goats into bioreactors for a precious protein usually begin with a study to determine whether it can be produced in mice, the animals with which scientists have the most experience in integrating "transgenes." The first transgenic mice are a kind of pilot plant. If one can
harvest the gene product of interest from them, it justifies the decision to
scale up to a larger animal that can produce it in large quantities. No human gene seems too large or complex for the system. In 1994 Pharmaceutical Proteins Limited (PPL) of Edinburgh, the commercial sister of the
Roslin Institute, the research group that cloned the sheep named Dolly
(see Chapter 23) and which pioneered the creation of protein-yielding
livestock, announced that in partnership with scientists at Zymogenetics,
Inc. (Seattle) it had created a mouse that makes milk loaded with fibrinogen, a key protein in the blood-clotting cascade. Fibrinogen is composed
of six interconnected protein chains that are made by three different genes;
one copy of it weighs the same as 340,000 hydrogen atoms. In an even
more impressive tour de force, two California biotech companies, Cell
Genesys and GenPharm International, created mice that make human
monoclonal antibodies, work which could lead to an important new set of
weapons to fight cancer. Scientists have even found genes for enzymes
called furins that perform the final touches (called posttranslational modification) in protein manufacture, such as adding sugar molecules, and
have succeeded in inserting them into the embryos of goats as well. The
presence of these genes in the goat will make the proteins in the goat milk
of even higher quality.

Since the arrival of Dolly, there have been astounding advances in
mammalian cloning. Among them, Genzyme Transgenics recently announced the birth of three goats, each of which is a clone of the same
transgenic goat embryo and all of which contain the human gene for

antithrombin III. This was accomplished by taking cells from a female goat embryo already known to carry the human AtIII gene and transferring the nucleus of each into a separate enucleated goat egg. The eggs are then transferred to surrogate mother goats that have been hormonally readied for pregnancy. Of 112 cloned embryos placed into 38 surrogates, three clones were born. This may seem low, but it represents a definite advance in the efficiency with which animals can be used as bioreactors (Dolly was the only success in 277 attempts). By cloning a female embryo, the researchers will of course produce only female offspring, all capable of producing milk with the human AtIII. Sandra Nusinoff Lehrman, who is the current CEO of Genzyme Transgenics, estimates that more than $200 million is spent in Europe each year to purchase AtIII.

In March of 2000, PPL Therapeutics, the biotech firm in Scotland that helped clone Dolly, announced that it had produced its first litter of cloned pigs. This is a key step in the effort to create pigs that have a human gene which will insulate their kidneys from rejection if transplanted into humans (Chapter 16).

Such advances are provoking protests, however. Until Dolly, the most famous farm animal in the world was Herman, the first transgenic goat in Europe, created in 1990 by a company in The Netherlands called "Pharming." At a 1997 scientific meeting, George Van Beynum, one of the company's executives, contrasted the generally positive response of the American public to the arrival of such animals with the deep uneasiness in his home country. The Netherlands has a law making it a crime to create transgenic animals unless one has first obtained a permit from the Minister of Agriculture, a permit which, due to political pressure, is almost impossible to obtain. Sustained pressure from animal rights groups in Denmark, Sweden, Germany, and Austria could lead those nations to adopt similar rules, which will make it much harder for their scientists to compete.

In the United States, one development that has enraged some animal rights groups is the practice of awarding patents on transgenic animals. One of the most famous patents in history is for the "Harvard mouse," a transgenic animal created in the laboratory of the renowned geneticist, Philip Leder, that carries in its genome a human gene with mutations that make it highly likely to develop breast cancer. Although the U.S. Patent Office had been issuing patents on inbred strains of plants and animals for decades, the award of the patent to Harvard in 1987 for a mouse engineered to carry a human gene struck some people as a new level of arrogance in the history of our domination of other species. In Europe during the mid-1990s, groups skeptical of genetics nearly won a battle over the language of the proposed European patent directive that could have made it much more difficult to secure the intellectual property rights to new products made by manipulating DNA. This would have surely been a pyrrhic victory, as it would seriously hamper the prospects for biotechnology in those countries.

MICE

Since I have been discussing the contributions of genetically altered animals to human welfare, I would be remiss here not to speak a few words of well-deserved praise on behalf of mice (*Mus musculus*). They are small, hearty, easy and inexpensive to care for, and in many ways remarkably like people. Scientists have been studying every aspect of their lives, from genes to courtship behavior, for decades. Because we already know them so well, powerful new tools in molecular biology promise to let us know them much better still. When the Human Genome Project, the federally funded effort to sequence all 3,000,000,000 DNA letters that make up the human blueprint, was launched in 1990, the planners knew that it was essential to develop in parallel equivalent efforts to learn the genetic maps and fine structures of other organisms. They chose the mouse. Among higher organisms, the level of detail of the genetic map of the mouse is today second only to that of humans.

Despite the fact that humans and mice have to go back more than 50 million years to find an ancestor that they share in common on the evolutionary tree, they are, comparatively speaking, cousins. Humans and mice have far more similarities than differences in their genes. This is because once nature has worked out an efficient solution to a challenging problem in cell physiology, it tends to reuse it in new organisms. Through the eons only modest changes drift into the DNA, and many of them are neutral in impact. That is, they do not really change the function of the protein for which they code. In two widely divergent species the protein differences may be no greater than say the performance differences between two comparably priced four-door sedans both using the same engine, but manufactured and marketed under different names.

Although we so far know the sequence of only a small fraction of the DNA in the mouse and human genomes, we can say with confidence that coding sequences are identical for about 80% of the base pairs. In most instances, the order in which genes sit on chromosomes is also the same in humans as in mice. Over the years, this has allowed scientists to compare gene locations in the two organisms and conclude that, for example, mouse chromosome 16 has many of the same genes lined up in the same order as does human chromosome 21.

Human and mouse gene mapping are in full swing at hundreds of labs around the world, and new data are poured into publicly accessible (via the World Wide Web) databanks every day. One of the first questions that a scientist now asks when he or she finds a gene of unknown function is whether or not a structurally comparable gene exists in other species. The first stop is often a computerized search of the mouse genome database. Just by finding that the mouse has a comparable gene suggests to the human geneticist that both are involved in some basic biochemical function. Often, the human geneticist may discover that a lot of hard work has been done for him. If a mouse geneticist has figured out what the gene does, it is very likely that in humans the comparable gene has the same duties.

With a powerful tool called site-directed mutagenesis, researchers can now deliberately destroy specific genes in mouse embryos in an effort to develop models of human diseases. These "knock-out" mice are today playing a key role in our pursuit of the causes of disorders ranging from breast cancer to Alzheimer disease. One of the first knock-out mice was designed to mimic Tay-Sachs disease, a fatal brain disorder of childhood caused by mutations in a gene that codes for a protein known as hexosaminidase A, which normally breaks down waste molecules inside the cell structures known as the lysosomes. In 1994 Richard Proia, a researcher at the National Institute of Diabetes and Digestive and Kidney Diseases at the NIH, crippled the comparable genes in mice embryos. As hoped, the mice grew up to have abnormal lysosomes, which makes them useful models to study the way the waste molecules accumulate in the brain. Oddly enough, however, these mice are relatively healthy. A few other examples of engineered mice include mice that have been created lacking a gene critical to the health of the placenta (they survive because healthy cells are transferred into the embryo at a key point to save a failing placenta), mice that lack a gene for a protein involved in the structure of myelin (which insulates nerves), mice that provide a model for multiple sclerosis, mice that lack one copy of the murine equivalent of the Huntington disease gene (and which have a reduced number of cells in the same parts of their brains as do human patients), and mice that lack the p16 gene, a condition which makes them susceptible to cancer. In one experiment, 9 out of 13 had spontaneously developed cancer by 26 weeks of age. The Alzheimer knock-out mouse was created by deleting the mouse version of a gene called presenilin-1 and inserting a mutated form of the analog human gene that is a cause of a rare hereditary form of the disease into mouse embryos. Scientists at several medical schools are working to develop other mouse mimics of Alzheimer disease. Together these animals permit a level of study undreamed of as recently as the 1980s. So the next time you see a mouse in your basement, think twice before you reach for a weapon. The quirky little guy has some cousins who might be working to save your life!

The Florida panther, highly inbred and nearly extinct, is being saved by the intro-duction of a closely related variety from Texas with which it can breed. (Photo © Larry W. Richardson.)

Endangered Species
New Genes Beat Extinction

No one knows how many species share the planet with us. E. O. Wilson, the great Harvard biologist and among the most qualified people on earth to make the estimate, has guessed that there are between 10 and 100 million. In his words, "We don't know, not even to the nearest order of magnitude." Wilson and several other biologists who specialize in the subject recently estimated that mankind knew of about 1.4 million species. Thus, by the reckoning of the some of the world's most eminent biologists, humans are only dimly aware of between 1 and 10% of the current cast of species on earth. We know virtually nothing about most of them.

Why? There are two reasons. Many live and die in places that most humans rarely even think about: the soil, the ocean depths, and the forest canopy, especially the canopies of the rapidly dwindling rain forests. As living creatures go, humans are big, and we tend to think mostly about the other big species that we routinely interact with—farm animals, the garden plants and song birds about our homes, the fish we catch. These constitute a tiny fraction of the total. The most enthusiastic amateur naturalist can probably identify fewer than 1000 species. But the vast majority of species, whatever their true number, are exceedingly small. Of the 1.4 million known species, more than half are insects, spiders, centipedes, or related organisms with jointed exoskeletons—hardly a favorite group among humans. The next largest group is the flowering plants. Of the 90–99% of the species that we may not even be aware of, the vast majority are microorganisms, mostly a riot of bacteria that are invisible to us.

Are they important? Unquestionably. Earth's great life cycles depend critically on them. Without insects to pollinate them, most flowering plants would rapidly become extinct. Bacteria and other soil organisms are nature's recyclers; in decomposing the dead of larger species, they reconstitute the topsoil that is constantly eroding. Without them there would be mass famine. Ironically, it is from among the handful of bacteria which we do know well that we have developed many of the antibiotics we use to subdue the otherwise often fatal infections caused by other bacteria.

There is good reason to think that, as man's expanding numbers consume the rain forests and pollute the waters, we are killing off species surely by the thousands, perhaps by the tens of thousands, every year. The logic that underlies this assertion is simple. Careful study of rain forest canopies, for example, yields new species on almost every expedition. For the most part, these are insects that over millions of years have worked out a little niche for themselves in an exceedingly complicated environment. Often it turns out that a newly discovered insect can survive only in a particular part of a particular species of tree because that is the niche in which its source of food evolved. Destroy the trees to make way for more fields for cows to support more human families and the organisms for which that species of tree is a world unto itself will perish.

Because they can see insects, committed field biologists and good amateur naturalists will continue to improve our inventory of them. For smaller species, about the importance of which to our own survival we know virtually nothing, it is highly likely that we will remain ignorant. We know of about 1000 viruses and 5000 bacteria, numbers that may constitute less than 1% of the actual diversity. Many of them may live in the guts of the insects that perish when third-world farmers struggling to feed their families cut down the last of a variety of a particular species of tree. They become extinct without ever having been recorded on mankind's global inventory.

What can we do to save the species which we know are endangered? The best approach is of course not to let them become so. The surest way to do that is to set aside large preserves, tracts of sufficient size that the larger species will have a fighting chance to survive. George Schaller, one of the world's great field biologists, has accomplished undreamed-of success in this regard. Among his many campaigns to influence governments to protect ecosystems, the greatest victory has been in China, which recently decided to protect an area in Mongolia that is nearly half as big as the eastern United States from human development. But many large species have recently become extinct, and many more are hanging from an evolutionary precipice. For them molecular genetics and reproductive biology may be the only rescue net.

Sexing the Spix's Macaw

What is the most endangered large animal species on earth? Possibly the Spix's macaw (*Cyanopsitta spixii*), a brilliantly blue colored, parrot-like bird native to Brazil. In early 1995 there were only 32 on the planet, most in the collections of exotic bird fanciers, and only one known in the wild. The Spix's macaw is sexually monomorphic; that is, males and females look alike. In the hope of bringing the species back from the edge of extinction, conservationists wanted to provide a mate for the last wild bird, but they did not know whether it was male or female. Ornithologists who had closely observed the wild bird's behavior thought it was a male, but could not be sure. Because it is the only remaining member of its species that has grown up in the wild, conservationists were extremely reluctant to risk harming it by capturing it for sexing. Molecular biologists provided a solution.

To answer the question, Richard Griffiths and Bela Tiwari, who work in the Department of Zoology at Oxford University, developed a molecular method to sex birds from the tiny amounts of DNA that can be extracted from the tips of molted feathers. In contrast to humans, in birds it is the females that have different sex chromosomes. In humans the female has two X chromosomes and the male has an X and a Y; in birds the females have one W and one Z chromosome and the males have two Z chromosomes. In 1993 Griffiths and Tiwari isolated a highly conserved gene (called C-W) which seems to be found on the W chromosome in virtually all but one group of bird species. They also knew of a second, closely related bird gene called C-2 that is located on a non-sex chromosome.

To develop a DNA test to sex the last wild Spix's macaw, they had to learn more about the nature of the C-W gene in macaws. To help them, a company called Stratagene made a genomic library from the DNA of a close relative, the hyacinth macaw, which is not endangered. A genomic library is simply a collection of bacterial colonies, each of which includes within it a little bit of the chopped-up DNA of an organism's entire set of genes. The bacterial colonies act as tiny, locked bookcases that protect the information until the researcher wants it. Unlike regular libraries, the books in genomic libraries are created without a card catalog. Fortunately, the molecular biologist has ways to rapidly scan the library to find the book (gene) of interest.

Griffiths and Tiwari used their chicken C-W gene as a probe to find its equivalent in the hyacinth macaw DNA. They next amplified 104-base-pair stretches of DNA from the C-W and C-2 genes in the Spix's macaw. There are just a few nucleotide differences between the sequences, but it is enough so that they can be easily distinguished. The C-W sequence includes a spot that can readily be cut by a special enzyme into two pieces of different length. The Z chromosome has no comparable sequence. Because both male and female Spix's macaws have two copies of the C-2 DNA which is not on either W or Z, it should be present regardless of sex. But only a female will have C-W DNA; a male, which does not have a W chromosome, will not show evidence of the C-W gene.

It took two years for field biologists to collect just three feathers molted by the wild Spix's macaw. Griffiths and Tiwari extracted DNA from the tips of these hard-won feathers and compared it to the DNA from five birds of known sex that are in captivity. All the tests on the known males yielded a single DNA band, as did the test on the wild bird. It is a male.

In the hope that this would be the result, conservationists had for five years been training a captive female Spix's macaw to live in the wild, even giving it lessons in how to find and eat natural foods. Late in 1995, they released her into the jungle less than 100 yards from her potential mate. I wish the story had a happy ending, but it does not. Less than a week later the female disappeared, and is presumed dead. The one encouraging development is that the local people, who over the decades destroyed much of the Spix's macaw's habitat, have been moved by the campaign to save the bird from extinction, and are working with the conservationists. Thanks to captive breeding programs, there are now 39 birds, and more releases into the wild are planned.

DNA Testing in the Zoo

Many of the world's major zoos now turn to molecular genetics to help them reassess breeding programs to preserve endangered species. This sometimes leads to surprising findings and bitter controversy. Why? It is difficult and expensive to acquire the founder stock, and if a zoo purchases an animal that is later shown not to be what it was represented to be, it will not be able to use it. In 1995 Dr. Terry Maples, director of Zoo Atlanta, stopped a plan to breed a pair of rare Sumatran tigers when tests suggested that the big cat he had purchased was probably the offspring of a cross in the wild between a Sumatran and the nonendangered Bengal tiger. At a 1995 meeting of the American Association for the Advancement of Science, Maples opined that the Sumatran cat had been "polluted" by Bengal genes. What then is the big cat that is the offspring of such a mating? According to the standard view among evolutionary biologists, if the Sumatran and the Bengal freely (even if infrequently) breed in the wild, they are members of the same species. Why then should Dr. Maples have halted the breeding program? Who decides the boundaries that define species and what constitutes a subspecies (or local variety) that is worth conserving?

India is home to dwindling populations of both Asiatic lions and Siberian tigers. Recent study of the DNA of 28 of the last 350 lions on the Indian subcontinent (many of which live in the Gir National Park in Gujarat) brought good news. The population appears to be sufficiently genetically diverse to have a decent chance of survival. DNA analysis of two wild Indian tigers brought less welcome news. They were found to carry gene variants usually only seen in Siberian tigers, which means that those two "species" now mate in the wild. Since there are many more Siberian tigers, in a few generations the wild Indian tiger will likely be superceded by a form that is much closer to the Siberian tiger, and a few decades hence, there will be no Indian tigers.

About 10,000 orangutans exist in the forests of Sumatra and Borneo, two large islands in Indonesia separated at one point by only about 50 miles of ocean. In Malay the word "orangutan" means "people of the forest." This reflects millennia of close contact. Until a century ago, the Indonesian people thought of these apes as primitive human relatives, a folklore that is easily understandable, especially if one spends time observing orangutans. It is not just that they look a lot like us; they also have behaviors like ours. They are smart too, capable, for example, of learning rudimentary sign language. Primatologists claim that they have taught individual orangutans to perform tasks as complex as picking locks, rowing boats, and making pancakes.

Fearful of the steady loss of habitat caused by the burgeoning human population, about 20 years ago the Indonesian government and a consortium of zoos in the United States set up crash breeding programs. At the time it was thought that orangutans from Borneo differed only slightly from those in Sumatra, that at most the two should be considered a pair of subspecies, and that forced crosses between them would be acceptable. This was largely because the two islands are thought to have been a single land mass until less than 100,000 (perhaps just 20,000) years ago. Conservation zoologists reasoned that not enough time had passed for them to evolve many distinct genetic differences. Furthermore, there was good evidence that for centuries humans had been transporting young orangutans, often as pets, between the islands, which suggests that the two groups had continued to mate until the quite recent past.

In the United States today there are about 80 orangutans that are the offspring of a Sumatran and a Bornean parent. All have been sterilized or placed on long-term birth control. Comparative DNA analysis of orangutans from each island indicates that they are much more different at the genetic level than was thought on the basis of comparing physical and behavioral differences. The most experienced observers can determine to which island group the adult males belong, and no one can correctly assign adult females. The Bornean males tend to have rounder faces and darker fur; the Sumatran males tend to have hair that is redder and more curly. Genetically, the differences are more dramatic. In one subspecies part of the chromosome number two (in each species chromosomes are arbitrarily numbered according to length, with the largest being designated as number one) is inverted compared to that of the other. Chromosomal inversions, in which a big chunk of DNA breaks off, flips, and rejoins the rest of the chromosome, are relatively common. Among closely related species, examining the differences in the number and shape of chromosomes, because they can be discerned under the microscope, was one of the earliest methods used to get some indication of evolutionary divergence. At the chromosomal level, the two subspecies of orangutan are more different from each other than are lions from tigers. This, and the fact that careful study of key proteins has discovered significant differences in the sequence of amino acids of which they are composed, has led many scientists to argue that the orangutans constitute two distinct species.

By painstakingly comparing samples of mitochondrial DNA from the two groups, scientists have concluded that the sequence differences are sufficiently great to suggest that they diverged at least 20,000, and possibly hundreds of thousands of, years ago. If correct, the longer estimate is crucially important, for it suggests that the two groups became reproductively isolated long before they were physically isolated by an ocean. This means that other forces had to be at work and, if true, is a powerful argument in favor of two distinct species. But opinion is not unanimous. One scientist, Dr. C. Cam Muir, a geneticist at Simon Fraser University in British Columbia, who studied the precise nucleotide sequence of five orangutan genes, concluded that the two groups were much too closely related to be called different species.

We are left with an odd situation. Over the last 30 years, we have decided that Sumatran and Bornean orangutans were no more different from each other than say Japanese are from Swedes. Realizing that they were on the brink of extinction in the wild, zoo programs have mated orangutans which are the descendants of animals born on different islands to each other. Now on the basis of DNA studies, some zoologists have decided that they have created a hybrid which, if it ever existed in nature, never took hold, and have decided that it should not reproduce. A few scientists have argued that this is a form of "primate racism" that recalls the 19th century obsession with creating a typology of humans in which northern Europeans were awarded the top spot. At the least, the sterilization of orangutans seems to be a bizarre feature of an effort to prevent extinction.

The fact is that there is no definitive way to say that two closely related varieties belong to one or two species. In the wild, the key criterion is reproductive behavior. With what animal does the animal in question mate? Even here, the edges are fuzzy. Depending on the situation, an animal that almost always mates with its own kind may drift just a bit to mate with a member of a group that is very closely related (as in the case of the Indian tigers). The current obsession with defining the perimeter of the orangutan species is by its very nature an arbitrary, some might argue foolish, exercise. The fact is that we know very little about orangutan DNA, surely much less than we know about their bodies and their behaviors. We must be a bit wary of making too much of DNA differences, which may turn out sometimes to mean less than we currently think. One wonders if Atjeh, the 33-year-old hybrid orangutan at the National Zoo in Washington, will someday be able to get his vasectomy reversed.

SAVING THE FLORIDA PANTHER

The Florida panther (*P. concolor coryi*) may be the world's most endangered subspecies of big cat. Surrounded by millions of their only predator, humans, by 1967 there were only about 30 of these big, gray-tinted cats teetering on the edge of extinction in the Big Cypress National Preserve in south Florida. For many generations, virtually all of the cubs have been born to matings between genetically close relatives, often siblings. To any veterinarian who examines one of these animals, the harmful consequences of inbreeding are immediately obvious. In addition to innocuous features like kinked tails and cowlicks, nearly all the adult males have cryptorchidism, an inherited defect causing one or both testicles to fail to descend into the scrotum, which reduces their fertility. About 80% of the cubs have significant heart murmurs, and in recent years three of them have died of atrial septal defects (holes in the wall that separates the two upper chambers of the heart). Virtually all the Florida panthers are badly infected with parasites, probably the result of a genetically weakened immune system.

The plight of the Florida panther is even more desperate than that of the orangutan. In 1992 the World Conservation Union warned that unless there was a sustained effort to repopulate the species, it would be extinct by 2055. In response, in 1995 the U.S. Department of the Interior, working in conjunction with state authorities in Texas and the Florida Fish and Wildlife Conservation Commission, captured eight female Texas cougars and resettled them in the Big Cypress Swamp. Zoologists chose this closely related subspecies because until about 150 years ago its natural habitat overlapped that of the Florida panther. If they found mates among the remaining male Florida panthers, the offspring could constitute a desperately needed dose of new genes for the highly inbred group. Less inbred animals would not enter life with two genetic strikes against them.

At first the rescue operation looked like it would fail. Three of the eight females uprooted from the arid climate in West Texas died. But the remaining five did successfully mate, producing 17 healthy kittens. Many of them have now reproduced, in some cases with a Texas cougar and in other instances with a hybrid offspring from one of the other four litters. In late 1999 zoologists estimated that the population of panthers in the Big Cypress Swamp had grown to 50–70 adults and kittens, and it looked like a robust breeding population would soon be reestablished.

Once again, evolutionary purists have begun to argue that if these two subspecies breed well and expand the population of pumas, the program will not have saved the Florida panthers. Instead, it will have created yet a new variety of big cat in south Florida. In the words of David Maehr, an assistant professor of forestry at the University of Kentucky who once led the recovery effort, the crossbreeding "is spiraling out of control." He thinks that if it continues for even a few more years, the distinctive physical features of the Florida panther will vanish. Dr. Stephen O'Brien, one of the world's authorities on the genetics of cat species, strongly disagrees. He has shown that throughout the Americas the various subspecies of puma (known regionally as mountain lions, cougars, catamounts, and panthers) interbreed. Thus, it is highly likely that until human encroachment trapped them in a small habitat, Florida panthers routinely mated with members of sister subspecies. Indeed, some of the Florida panther's distinctive features probably arose only after it was trapped in the Big Cypress Swamp.

Conservationists hope to rebuild the panther population in Florida to 500 animals and to increase its range. Not surprisingly, as a few of the big cats have expanded their perimeter, people living in the region have protested. If the population remains under 500 and geographically contained to its current region, to counter the devastating effects of inbreeding conservationists will have to import new animals about every decade.

The cheetah, the world's fastest land animal, probably once came as close to extinction as the Florida panther is today. DNA studies have yielded evidence that the species passed through an evolutionary bottleneck, which means that all of today's animals are closely related. Studies of genetic variation among the big cats indicate that they have less than 10% of the genetic diversity expected in a species. Cheetahs are so inbred that it is very likely that a skin graft from any one to any other will take, indicating immunological identity! It also means that the species is in constant peril of being wiped out by a viral or other infection. In fact, cheetahs recently had a close brush with the feline infectious peritonitis virus, an agent that causes only mild symptoms in a tiny fraction of the house cats it infects. In them the virus is deadly; all the infected animals became seriously ill, and 60% died. Unfortunately, there are few places on earth where one can attempt to restore a natural cheetah population in the manner that is being attempted with the panther in Florida. Major zoos have been trying to breed cheetahs in captivity, but have had little success. Sometime in the 21st century, we will have to launch breeding programs for the cheetah as intensive as those currently under way for the Florida panther or we will lose that wonderful animal from the planet.

The rescue of the whooping crane from the brink of extinction provides good evidence that the effort to save the Florida panther can succeed. In 1941 there were only 22 whooping cranes left in the world. Since that time continuous and caring intervention by man, largely through the work of the International Crane Foundation headquartered in Baraboo, Wisconsin, has brought the population to about 240. DNA testing will soon play an important part in the continual rebuilding of the whooping crane population. Conservationists plan to study the DNA of the young adults and arrange matings that maximize the chance for genetic diversity in offspring. This should result in higher survival rates and a more robust population.

Why take on the quixotic task of trying to save nearly extinct primates, cats, birds, or some of our even more distant relatives? If neither reverence for life nor an aesthetic appreciation of nature is enough, there is a much more powerful rationale with which to persuade the doubters who would prefer to pave paradise and put up a parking lot. The world's biota, the sum total of living things, contain within their DNA information that will, if preserved, studied, and understood, offer humans dazzling new knowledge about disease and powerful new medicines to fight the infections that have always and will always plague humans. Every time the evolutionary light winks out for a species as seemingly inconsequential as an insect with a restricted range in the Brazilian rain forest canopy, we may have lost our chance to find a wonderful agent in the fight against cancer or AIDS. Of the world's ten most prescribed pharmaceuticals, nine derive from natural sources.

The black plague killed off more than a quarter of the population in Europe in the 14th century. Until it gradually attenuated, syphilis killed millions of Europeans and Asians as it spread across the globe in the 16th century. Measles and smallpox, introduced into the Americas and Polynesia by explorers, also killed millions. In the 19th century, tuberculosis, the white plague, was the number one cause of death among adults on both sides of the Atlantic. Although we have no idea what they were, over the millennia untold numbers of plagues have wiped out huge fractions of various animal and plant species. My point is that within the genomes of all living species there are molecular secrets indicating why some individuals survived to reconstitute the population. These data can provide genetic information with which to meet similar new attacks in the future. A compelling example is the discovery that a small number of humans carry a variant gene that renders them highly resistant to developing AIDS. The HIV virus can infect them, but cannot cause the full-blown disease. Study of this gene may point us in an important new direction in drug development.

The pharmaceutical industry has long screened countless compounds from nature to find the few that will become important new drugs. They have not even scratched the surface of biodiversity. Although there is great current interest in developing an in-house approach (called combinatorial chemistry) to generating important new compounds to treat disease, it would be unwise to ignore the wealth still hidden in nature. Companies such as Shaman Pharmaceuticals, a small botanical prospecting company in California, have the right idea. Their scientists are trying to draw upon the knowledge of indigenous peoples about locally used medicines to find important new compounds.

There are also many secrets to be found by studying the genomes of organisms that one is trying to protect. A strain of wild mouse in California was recently found to be highly resistant to a viral infection that was devastating other strains. Closer study showed that at some point in the past, the resistant strain had successfully integrated a portion of the viral DNA into its genome, making it essentially immune. As many as 3,000,000 DNA base pairs in our genome may be the "genetic fossils" of viruses that have infected our species or the primate from which it arose over the last 10 million years. It may be that the secret to halting the spread of future plagues lies in the analysis of human DNA sequences to retrace our ten-million-year struggle with viruses and bacteria.

Five cloned piglets, Millie, Christa, Alexis, Carrel, and Dotcom, created by the sci-entists at PPL Therapeutics. (Courtesy of PPL Therapeutics Ltd.)

Xenotransplantation
Animal Organs to Save Humans

THE HARDEST WAIT

Every year in the United States and Europe, surgical transplant teams save ever more lives, yet every year the number of people who die while waiting for a donor heart, lung, liver, or kidney grows. Between 1990 and 1995 in the United States on average each year only 4,835 people became organ donors after their death, about 1 in 500 of all those who died. In most cases, the donors were young persons who died of traumatic head injury, and each could provide several organs. During 1994, about 5,000 newly dead persons provided organs for 18,000 transplant operations in the United States.

Despite all that surgery, in 1996 the official waiting list, which is maintained by the National Organ and Bone Marrow Registry, included 48,000 people who were in imminent need of a donor organ. Of these about 33,000 needed kidneys. And these numbers significantly underestimate the real need. Some cardiologists have guessed that as many as 50,000 people have sufficiently severe heart disease to qualify for transplant if a donor organ was available. To put it coldly, the people on these lists (and many more who are headed there) have little choice but to hope that fate will strike down someone else so that they can win a new lease on life. Each year in the United States more than 3,000 people who have become sick enough to claim a spot on a list die while waiting for a donor organ.

There is a severe shortage of cadaver organs in virtually every country that routinely performs transplant surgery. For example, during 1995 in Great Britain, 1,003 cadaver organs (including 36 imported from other nations) became available for surgical use. Yet, at the end of that year, there were 6,133 patients on the United Kingdom Transplant Support Service Authority's waiting list, a number that may underestimate the true need,

as the criteria for being placed on that list are more stringent than comparable rules in the United States. In 1995 only 3,679 persons died from traffic accidents in all of Great Britain, and only 5,235 (much less than a third
of the number in 1970) died of stroke (which is after trauma the second
major source of organs).

As we live longer and as physicians learn to manage chronic heart, kidney, liver, and lung diseases better, as deaths from stroke continue to drop,
as highway safety improves and alcohol-related traffic fatalities drop, and
as more cyclists wear helmets, we can be certain that the waiting list for
donor organs from cadavers will continue to grow.

How the Field Developed

The hope that we might someday be able to use animals as organ donors
is older than transplant surgery itself. Alexis Carrell, who worked at
Rockefeller University in New York and won the Nobel Prize in 1912
for his pioneering work in cell culture, spent years trying unsuccessfully
to transplant organs and limbs from one animal to the other. It is said
that his research inspired H. G. Wells to write *The Island of Doctor
Moreau,* the story of a brilliant surgeon in self-imposed exile on a Pacific island who has discovered how to create chimeras of animals and
humans.

With little knowledge to go on, but faced with desperate patients dying in front of them, in the years after World War II a small group of surgeons began studying organ transplantation, especially kidneys, in animals, often using dogs as their models. The first decade was devoted to
grasping the outlines of the problem of tissue incompatibility. Having
made modest advances, on a few occasions some research physicians attempted what, in retrospect, were operations as fantastic (and some said as
foolish) as those imagined by Wells. In 1964 an American surgeon performed the first xenotransplant (the term refers to any transplantation of
an organ from one species to another) when he removed a heart from a
chimpanzee and placed it in the chest of a man who was about to die from
heart failure. The much smaller chimp heart could not handle the circula-
tory load and the man died two hours later. In 1968 surgeons in the United
States attempted a sheep-to-human heart transplant, and surgeons in the
United Kingdom attempted a pig-to-human heart transplant. In both

cases the immune system immediately rejected the foreign organ, and the patients died within hours. In 1977 South African surgeons twice transplanted baboon hearts into dying men; in both cases the patients quickly rejected the hearts and died within a few days. In 1992 Polish surgeons transplanted a pig heart into a dying man, and the smaller organ kept him alive for 20 hours.

The most famous and controversial cardiac xenotransplant was attempted in 1984 when surgeons at Loma Linda University Medical Center in California put a baboon heart into a 15-day-old infant known as "Baby Fae." This desperate measure was defended as a bridging operation, an effort to keep the child, who had an undeveloped left ventricle, alive until a human donor organ became available. By using drugs to suppress the infant's immune system, the doctors were able to keep Baby Fae alive for 20 days. During the same era there were a few similar efforts to use baboon, chimpanzee, and pig kidneys and livers as xenotransplants. In a few cases patients survived for up to 10 weeks, but in no case was anything approaching normal organ function achieved.

Tricking the Immune System

While the occasional heroic, some might say quixotic, emergency xenotransplants were grabbing headlines, teams of immunologists took on the arduous task of trying to understand why recipients rejected the organs so quickly. Studies of xenotransplants from pigs to dogs revealed that the immune system of the recipient animal sometimes mounted an overwhelming attack on the donor organ that started within seconds, and often caused death in a few minutes (a response that is now well understood and called a hyperacute rejection or HAR). There seemed to be an insurmountable immunological wall that would make xenotransplants impossible. Nature had clearly never anticipated the need for xenotransplantation. Over the eons, mammals have evolved an elaborate set of genetically coded immunological defenses that will, when they recognize that the body has been penetrated by foreign tissue, unleash a massive counterattack so deadly that it carries an extremely high risk of destroying the patient.

Surgical advance in xenotransplantation depends on cracking the secrets of the immune system and then learning how to circumvent its defenses. Beginning with studies in the mouse and moving on to other species, immunologists have gradually characterized an immensely complex set of genes that make up what has become known as the major histocompatibility locus. This block of genes directs the production of molecules that reside on the surfaces of circulating cells that form part of the immune system. They immediately detect any foreign organ. In humans it is called the HLA (for human leukocyte antigen) system. By 1970 scientists had found that these genes (which reside on chromosome 6) have so many variations that in combination there are more than 20,000 possible genotypes. That is why when one searches for a donor outside one's immediate family the odds are so low of finding an immunologically identical match. The dramatic increase in successful kidney transplants in the 1960s and 1970s was in large part due to a better understanding of which genes in the HLA complex must be matched in order for the operation to succeed. It is our ability to characterize the HLA type of any individual that made it feasible to create a national donor matching program.

Over the last five decades the immense effort to understand our immune system has paid off. The first human kidney transplants, which were performed at the Peter Bent Brigham Hospital in Boston in the mid-1950s, involved identical twins. Surgeons and their medical colleagues gradually branched out to transplants from brother to sister and parent to child. That is, they began by transplanting organs between genetically identical individuals and progressed to genetically similar donor-recipient pairs. As they learned more and more about how to avoid organ failure due to the response elicited by foreign HLA antigens, physicians became able to match organs between unrelated persons, thus greatly expanding the pool of potential donors.

By 1980 we knew that by matching just four particular major HLA alleles, the odds of transplant success would substantially improve. Surgery was also greatly aided by immunosuppressive drugs, especially cyclosporin A, which became available about that time. Cyclosporin A is almost magical in its power to dampen the immune response, in effect letting us circumvent some of the problems associated with making imperfect matches between donor and recipient. Transplant teams now know that if they can find a fresh donor kidney which shares an immunological profile at the four key HLA coding sites with the potential recipient, the odds are 9 out of 10 that the kidney will be functioning well in the recipient two years after the surgery. For many recipients, the donor organ lasts much longer. Yet, with all this success the waiting list keeps growing and each year more people die while waiting for a transplant.

By the 1990s immunologists knew enough about how organisms react to foreign tissue to develop a visionary strategy to solve the tremendous shortage of organs. The first task was to pick a donor species and develop much deeper understanding of its immune system. Because they are far more plentiful than nonhuman primates and much less expensive to maintain, and because their hearts and kidneys are about the same size as ours, scientists chose pigs as the future source for our replacement organs. Researchers also still have much interest in baboons, which are the second most common primate (after humans) on the planet. Baboons, however, have posed difficult problems, not the least of which is that their blood types do not include a universal donor (like human type O) blood. Nevertheless, during 1994 Thomas Starzl, a pioneering transplant surgeon and researcher at the University of Pittsburgh who helped to develop cyclosporin A, transplanted baboon livers into two patients who were near death. His team was able to keep the patients alive 25 and 75 days. In effect they overcame the risk of the hyperacute rejection (HAR) that kills in minutes and the acute vascular rejection (in which the patient's immune cells attack the blood vessels of the donor organ) that kills in days, but could not defeat the acute cellular rejection mounted by the patient's T cells that develops over weeks.

The first problem to solve in either baboon- or pig-to-human transplants was to suppress the HAR, which is what so quickly killed the dogs that received the pig hearts in early surgical experiments. During the HAR, circulating cells and chemicals in the body immediately recognize the foreign organ and call an immunological "red alert." Within seconds, millions, then billions, of cells are triggering a massive counterattack. A group of proteins making up what is called the "complement cascade" almost immediately destroys the cells that line the blood vessels of the donated organ and in turn the organ itself. The patient dies from massive bleeding and anaphylactic shock.

The solution to circumventing the HAR in xenotransplants came with the ability to design transgenic animals. Two major approaches emerged: (1) manipulate early pig embryos to delete a gene that produces the key cell surface marker that human immune cells called xenoreactive antibodies immediately recognize and attack and (2) create transgenic pigs that carry one or more human genes which regulate the production of complement and therefore can moderate the power of the complement cascade. Early work suggests that both approaches might work, but there is much to be done. For example, mice that have been born from transgenic embryos in which the gene for the Gal epitope (which codes for a key cell surface antigen by that name that antibodies recognize as foreign and attack) has been deleted provide donor organs which recipient mice do not reject, but the donor mice are blind. Scientists have successfully created transgenic pigs that carry both monkey and human genes which code for proteins to regulate complement activity (RCAs). When kidneys from these pigs were transplanted into monkeys, the organs in some survived for up to 60 days, long after the time for acute rejection. Scientists in both universities and biotech companies have launched programs to create transgenic mice, pigs, and other animals that will carry several genes in addition to the RCAs. The hypothesis and hope is that animals that receive such donor organs should initially perceive them as not too different from their "immunologic selves," and that their immunological sentries will not call in an HAR.

Since 1993 David White, a surgical researcher in Cambridge, England, has been developing a herd of transgenic pigs that contain in their genomes multiple copies of a human gene called decay accelerating factor (DAF), the earliest acting of the proteins that are critical to the HAR. When White removed hearts from these transgenic pigs and perfused them with human blood, he found that the organs sustained much less tissue damage than did hearts from ordinary pigs. He and other scientists have transplanted transgenic pig hearts heterotopically (put them into animals that have kept their own hearts) into baboons (which have RCAs quite like humans). In experiments in England and in the United States, the transgenic pig hearts have delayed the HAR and lived up to 30 hours, about 30 times longer than might be the case if a full force HAR was unleashed. On the other hand, a few control (nontransgenic) pig hearts pumped nearly that long when placed in the chests of cynomolgus monkeys. When Dr. White's team gave the recipient monkeys drugs to suppress their immune systems, they were able to keep them alive with pig hearts for 60 days.

At a company in Boston called Biotransplant, Inc., a group of scientists led by Elliot Liebowitz, Ph.D., is developing an even more sophisticated, two-pronged strategy. In addition to breeding transgenic pigs with organs whose surface cells should be able to neutralize part of the human immune response, they are focusing on ways to alter the human recipient's bone marrow to make it pig-like. A technique that has worked in several animal species and could work in humans involves removing some bone marrow from the individual who needs the transplant, culturing the cells in laboratory flasks, using genetically engineered viruses to force the cells in culture to take up pig genes that code for key immune response proteins, and returning the altered marrow (by a simple intravenous catheter) to the patient. If it works in humans, the recipient would have an immune system that defended its body in the usual ways, except it would regard a pig organ as "self." In November of 1999, BioTransplant announced that, in association with physicians at the Massachusetts General Hospital, it had performed research with a small group of human subjects that provided proof in principle of this approach, which they call ImmunoCognance (TM). Their work, which was reported at the Fifth Congress on Xenotransplantation in Nagoya, Japan, triggered a $2,500,000 payment from Novartis, the giant pharmaceutical company that is funding much of the research.

Just as with human-to-human heart transplants in which success was measured at first by survival times of hours, then days, then weeks and months, and finally years and decades, success with porcine transplants will grow slowly. But grow it will. On the basis of cross-species animal-to-animal transplants, some immunologists think even now that it is scientifically permissible to try a pig-to-human heart transplant. They envision the attempt with a patient who has only hours to live and for whom no human donor heart is available. They think that by using drugs to suppress the human immune response they can prevent rejection of the transgenic pig heart for up to 30 days, enough time to justify a "bridging" operation—surgery to fend off death in hopes that a donor organ will become available. However, before that attempt is made, those who favor xenotransplants, a loose coalition that includes millions of patients and their families as well as physicians and scientists, will have to neutralize the objections of a perhaps even larger coalition of opponents, mostly animal rights activists who question both the safety and the ethics of the entire field.

THE SAFETY ISSUES AND THE ETHICAL DEBATE

The safety and ethical issues raised by xenotransplantation first captured general public interest in 1995 when a team of AIDS researchers in San Francisco boldly proposed transplanting baboon bone marrow into human patients with the disease. It had been known for some years that the bone marrow of baboons is highly resistant to HIV infection. Even when they are directly inoculated with massive quantities of HIV, baboons do not get AIDS. The idea behind this controversial experiment was that if transplanted baboon marrow cells took hold, the bone marrow of the infected patient would become a chimera—partly composed of his own cells and partly composed of those of the baboon. If the marrow made lots of baboon T cells that are adept at fighting HIV, they might even wipe out the virus, which would constitute the world's first true cure of AIDS.

As word about the proposed baboon-to-human transplant spread, some bioethicists argued that the experiment would be unethical because the operation posed an unquantifiable, but extremely high, risk of death to the human recipient (whose immune system was severely weakened) from infection with an endogenous baboon virus or from some more usual infectious agent. But public discussion quickly turned to the much more frightening question of whether such surgery could cause an epidemic—the result of the inadvertent transfer of some unknown baboon virus into humans. Because no one can design an experiment that offers absolute reassurance about this risk, it is a concern that is difficult to allay.

We cannot predict what will happen when a virus that has for millennia co-evolved with one host suddenly jumps into a new host, but we do know that it can be devastating. In 1967 vervet monkeys infected with the Marburg filovirus were imported into Germany. The virus infected 31 humans, and killed 7. In 1976 a shepherd in Pakistan developed a bleeding disorder that turned out to be due to a rare infection with the Crimean-Congo hemorrhagic fever virus, a bug previously thought to be unable to infect humans. Seventeen persons, mostly medical workers who had contact with the man, became infected, and four died. During 1995 in the Four Corners region of the American west there was an outbreak of hantavirus (a virus that is routinely found in rats but does not cause them any ill effects) in humans. About one-half of the infected persons died. Even more frightening is the possibility that the flu epidemic of 1918, an illness that killed 25,000,000 people around the world in a year, was caused by a mutated form of the swine flu virus.

Frightening as they are, acute outbreaks are much easier to diagnose and contain than are epidemics in which the virus causes a devastating illness that develops slowly over a much longer period, exactly what occurred with AIDS. There is strong circumstantial evidence that the HIV epidemic began quietly in Africa when a monkey virus known as SIV crossed into humans, mutated, and adapted to its new home, and then moved on to other humans. The virus infected millions of people during the decade before we realized that a global epidemic was under way.

When the FDA was asked to permit the baboon-to-human experiment, it held a special meeting to review safety issues. The focus of attention was the baboon foamy virus, thought to be harmless to humans, and whether passaging it through a human might select for mutant strains which could become highly virulent. After the meeting, which was held on July 14, 1995, the FDA demanded that the medical team make every possible effort to use a baboon that did not carry the virus. It also required that San Francisco General Hospital beef up its containment facilities. Despite uncertainties about the risk to the public health, in August the FDA gave approval for a clinical trial involving one patient.

On December 14, 1995, Jeff Getty, an AIDS patient and activist in San Francisco, became the first human to receive a transplant of bone marrow cells taken from a baboon. The experiment did not harm Getty, and there was no evidence that he developed a viral infection of baboon origin. However, within a few weeks it was clear that the stem cells did not engraft (establish a permanent colony in his bone marrow) so the procedure should not help him in his struggle with AIDS. How could the doctors determine this? They used PCR to search Getty's blood for DNA sequences that are unique to the baboon Y chromosome. The technique, which could find just one baboon cell among 1,000,000 human cells, found none. Ironically, one reason for the engraftment failure may be that the experiment was too tentative. The transplant team may not have been sufficiently aggressive in suppressing his weak, but functioning immune system, so it might still have been strong enough to reject the baboon cells.

Meanwhile, other experimental firsts that posed risks of transmitting animal viruses into humans were being announced at a rapid clip. In April, 1995, Diacrin, Inc., a company in Massachusetts, began a trial to investigate the safety of transplanting fetal pig brain cells into the brains of persons with advanced Parkinson disease or Huntington disease. The hope was that these immature cells would escape immune surveillance and grow to produce the dopamine that is so deficient in Parkinson patients and to replace the cells that die off in a region called the substantia nigra in Huntington patients. Scientists at Diacrin launched their human study only after showing that they had been able to deliver the fetal cells to the precise location in the brains of rats and demonstrate that the foreign cells lived and established connections with rat brain cells. Also in 1995, Duke University physicians tried to treat a patient with end-stage liver failure by passing his blood through a pig liver that was outside of, but hooked up to, his body, a living dialysis machine. Elsewhere, scientists started using encapsulated pig pancreas cells to treat severe diabetes. Others began transplanting adrenal cells from the fetal calf into the spinal cord of patients with end-stage cancer in an effort to relieve intractable pain. As word of cell xenotransplants spread, a chorus of critics both inside and outside the animal rights community emerged.

Knowing that it was their most likely means of reaching a wider constituency, animal rights groups argued both that using animals as donors was unethical and that pig xenotransplants posed a major public health risk. They warned that the porcine endogenous retrovirus (PERV), which is of the same general class as HIV and which is carried in the DNA of all pigs, could infect humans and become the next AIDS-like epidemic. In the United States, expert virologists were not worried. In 1996 the Institute of Medicine asked 60 virologists to advise it if xenotransplants posed a serious threat to humans; 59 of them said that the experimental work posed little threat and should proceed.

In England, at the time home to the world's leading xenotransplantation research, a strong coalition of animal rights groups argued that even if the research did not pose a public health threat, it constituted the "ultimate exploitation of animals." The groups had an obvious political impact. In 1996, as researchers talked more and more openly about performing the first animal-to-human organ transplant, the government imposed a moratorium and convened a blue ribbon commission to assess the ethics (not the safety) of xenotransplantation. About the same time, Dr. Karl-Frederick Bopp of the Health and Social Policy Division of the Council of Europe called for a moratorium throughout Europe on such work. Late in

1996 the British Advisory Group on the Ethics of Xenotransplantation advised that too little was known about the risks to ethically permit clinical trials in humans.

In 1997 U.S. officials at the FDA, confronted with mounting public concern about the risk of PERV and strong resistance from animal welfare groups, placed a clinical hold on all ten approved research programs that intended to transplant whole pig organs into humans. They required each to demonstrate the capacity to detect even the smallest signs of PERV infection in recipients. Within six months, six programs had successfully done so and had restarted. In June of 1999 the FDA Xenotransplantation Subcommittee held a public meeting as it prepared to issue final rules on the conduct of xenotransplantation research. Expert after expert testified about efforts to detect evidence of PERV infection in persons who had been exposed to porcine tissue. Together they had tested every human research subject who had received pig tissue and had found no sign of infection in anyone. In addition, they had tested 100 primates that had received pig tissue, and none of them was infected with PERV. The meeting was notable for the report that in the first half of 1999 more than 150 pig-to-primate transplants had been conducted. At the close of the meeting, Dr. Hugh Auchincloss, chair of the FDA subcommittee, announced that the FDA would give a green light to cell xenografts (experimental therapies such as using fetal pig neurons to treat Parkinson disease). The conditions under which the FDA will permit xenotransplants will be published in 2000.

In Europe, the many animal rights groups who abhor the use of transgenic animals as spare organ factories for humans have become allies with religious groups that are troubled by the notion of transplanting animal organs into humans. Some of them believe that deep spiritual consequences flow from placing an animal heart into a human; others that such actions violate the natural order. Others worry that it would be psychologically devastating for a person to depend on an animal organ for existence.

How do people feel about the possibility of living with a baboon, pig, or some other genetically engineered animal organ in their bodies? We have only a little evidence. A survey of 1728 acute-care nurses in Australian hospitals found that more than 65% said that they would not accept a primate organ and only 19% reported that they would. The percentages were virtually identical when the same question was asked concerning pig or sheep organs. The survey did not ask whether views would differ if the donor organ contained a few human transgenes to modulate the immune response. On the other hand, Jeff Getty was eager to undergo the baboon bone marrow transplant. A survey of healthy persons may offer little insight into what most gravely ill persons would do should xenotransplantation represent the only option to avert death.

Many persons alive today because of a heart transplant tell of feeling a profound connection with the donors that they never met. Some even report feeling that the deceased individual is watching over them, almost invariably in a kind, caring way. Given the symbolism of the heart in our culture, the use of transgenic hearts, be they taken from primates or pigs, is likely to pose significant emotional issues for some recipients. How would it feel to wake up each day to listen to the beat of a baboon's heart? The question is impossible to answer, but one prediction can be confidently made. The early recipients will have to bear the burden of intense media attention. They will be forever redefined by others as the men and women who first had animal hearts placed into their chests.

Hopefully, they will benefit from the fact that we live in an era that has recognized and embraced a neurological definition of self. For most of our cultural history the heart has been regarded as the seat of emotional life. For 2000 years we have defined death as the moment at which the heart stopped. Paradoxically, it was the possibility of human-to-human heart transplants that moved technologically advanced societies to adopt a neurological definition of death. This was legally necessary to permit medical teams to harvest viable organs from brain-dead individuals for donation.

Despite how technologically advanced it may seem to save human lives with genetically engineered pig hearts, the use of transgenic organs to provide replacement organs will probably last only a decade or two. It will be superseded 20–30 years from now by the creation on demand of needed organs grown by cloning and reprogramming cells taken from the needy patient. Scientists are already trying to develop ways to clone organs from single cells. Needless to say, this scenario suggests that those who could afford access to such organs might be able to dramatically extend their life span. This may be the greatest legacy of an era that has just begun with the cloning of a sheep from a single cell.

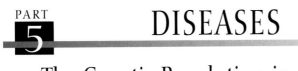

DISEASES

The Genetic Revolution in Medicine

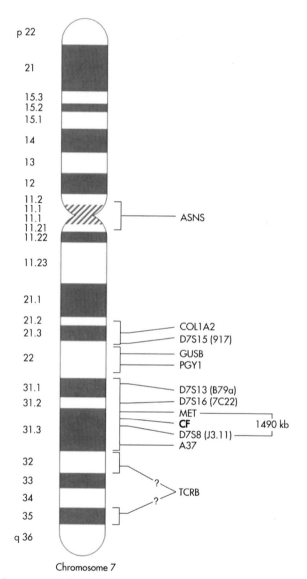

Schematic chromosome showing DNA probes used to map the location of the Cystic Fibrosis (CF) gene. *(Reprinted, with permission, from Watson et al. 1992 © Scientific Books/W. H. Freeman and Company.)*

Cystic Fibrosis
Should Everyone Be Tested?

The Long Road to the Cystic Fibrosis Gene

When I first became interested in human genetics in the early 1970s, most persons with cystic fibrosis (CF) died before their 20th birthday, many at a much younger age. For the more severely affected children, CF was a parent's worst nightmare. For them childhood unfolded in a world of chronic illness. As lung function deteriorated, each hospitalization became a battle to snatch a beloved child from the hands of death, and each victory was temporized by the certainty that another battle loomed ahead. The losses were horrible. One of my most painful memories of medical school occurred during my rotation in pediatrics. When his 12-year-old daughter with cystic fibrosis died of pneumonia in the intensive care unit, a father tried to commit suicide in the hospital.

As recently as 1974, a leading medical text described CF as "not only one of the most common disorders of childhood, but... also one of the most enigmatic." Even its name reflects the mystery. The term "cystic fibrosis" was coined more than 60 years ago to describe the scarring found in the pancreas at autopsy, yet the far more devastating effects are on the lungs, in which the ravages of recurrent, serious pneumonia are evident at a glance. Most strange is the protean nature of the disease. Some children are severely ill from birth, others become ill only gradually over four or five years, and a few are quite well until young adulthood. Most patients have both lung disease and severe gastrointestinal problems, but some are burdened only with comparatively mild lung disease.

Scientists demonstrated that CF was an autosomal recessive disorder (the result of being born with two defective copies of the same gene, which if present in only one copy does not cause illness) in 1949, but they remained ignorant of the cause for 20 years. In 1967 a researcher showed that the serum of persons with CF contains a chemical factor that inhibits the movement of the tiny cilia in rabbit trachea. This suggested that the ravages of CF might arise from a defect in how some factor flowed into or out of lung cells. By the mid-1970s, scientists had compiled enough evidence to be confident that the injuries caused by CF could all be explained by a defect in the transport of ions across cell membranes, but they did not know what caused this to occur.

To a geneticist, one of the strangest aspects of CF is that it is such a common disorder. About 1 in 2500 white children is born with CF (the disease is less common in other racial groups). That means that about 1 in every 25 whites carries a single copy of the CF-causing allele. The odds of 2 carriers marrying is 1/25 × 1/25 which is 1/625. The chance that an at-risk couple will bear an affected child is 1 in 4 with each pregnancy. 1/625 × 1/4 equals 1/2500. But how could such a "bad" allele become so common, especially since persons with CF virtually never have children (affected men are infertile and affected women are often not healthy enough to risk a pregnancy)?

One of the first appealing hypotheses was proposed 30 years ago. Dr. Alfred Knudsen, who now works at the Fox Chase Cancer Center in Philadelphia, calculated that the unusually high prevalence of the CF allele could be explained if healthy persons who carried one copy of it were on average just 2% more fertile than were those who did not. That would be enough to maintain the allele at high levels despite the deaths of affected children. What aspect of the CF allele might confer a reproductive advantage on carriers? There have been several creative guesses, such as that by a scientist who suggested that in women who are carriers the vaginal fluids might have a different viscosity that would facilitate the movement and survival of sperm, increasing the chances of becoming pregnant. Recently, studies in India have shown that CF carriers are more likely than are non-carriers to survive cholera. Since cholera has been sweeping through human populations for millennia (it devastated the United States in 1832, 1849, and 1866), this is a plausible (though still unproved) explanation for the high frequency of the CF allele.

When gene mapping tools became available in the early 1980s, many top research teams took up the challenge of finding the CF gene. They studied the families burdened with CF, seeking to identify which of their ever-growing number of DNA markers were present in children who had the disease and absent in the brothers and sisters who had not inherited the disease. The DNA markers that met that test would have to be near the hidden CF gene. In 1985 a group that included academic and commercially based research teams announced that they had used a linkage approach to localize the CF gene to the long arm of chromosome 7. The CF gene was the first to be mapped by what was for a time called "reverse genetics" (now called "positional cloning"). The term was meant to reflect the fact that such studies permitted researchers to find genes even when they did not have a clue as to their function. In less than a year, other groups had found markers on each side of the gene, narrowing the area on chromosome 7 where it had to lie. Even with this exciting advance, which led to the first prenatal test for cystic fibrosis, the road to elucidating the CF gene remained long and arduous. It took several major scientific groups nearly four years to hunt through the 1,490,000 base pairs of DNA in that region of chromosome 7 until one team won the race in the fall of 1989. One of the leaders of the team that cloned the CF gene was Francis Collins, then a young professor of human genetics at the University of Michigan. This discovery propelled him to his current position as Director of the National Human Genome Research Institute.

The cloning of the CF gene and the elucidation of the complete amino acid sequence of the protein for which it codes was a gigantic leap for molecular medicine. By studying the protein's structure and location, scientists were able to show definitively that its primary function was to transport chloride ions across cell membranes, thus confirming the long-held hypothesis. It was named the CFTR (cystic fibrosis transmembrane conductance regulator) gene. This discovery raised exciting new approaches for research. For example, by testing patients and studying the correlation between various mutations that occur at different sites on the gene with the severity of the disease, scientists began to understand how CF could express itself in such varied ways.

Among whites of northern European ancestry, about 70% of all CF alleles are missing the same three base pairs of DNA, causing the protein to lack just 1 amino acid (a phenylalanine at position 508) out of more than 1000. Because this mutation is so common, it was hoped at first that just a few more mutations would account for the other 30%, which would make it easy to develop clinical tests to identify persons who carried the CF gene. But the CF gene is large; its coding region contains 24 different regions (exons) and uses more than 6000 base pairs of DNA. We now know that there are hundreds of different possible mutations, some found only in a single family (nicknamed "private" mutations). This made the task of developing a clinical test to identify carriers more difficult, especially for people of southern European extraction, a group in which the 508 deletion accounts for only about 30% of all CF alleles. It also greatly complicated the job of correlating the severity of disease with the pair of mutations (since each parent could contribute any one of hundreds) with which a patient was born.

The chore of correlating genotype with phenotype, a task that must be repeated every time investigators find a gene that contributes to a disease, has already taught some important lessons. In delineating the CF gene, geneticists, who had been lumpers, became splitters. For example, they found some adults with uncommon mutations who had such mild symptoms of lung disease that no physician had ever guessed they had CF. Even more astounding was the discovery that among otherwise healthy infertile men there are a small number who have mutations in each of their CF genes. Undiagnosed cystic fibrosis may account for 1–2% of male infertility. Such discoveries greatly enlarged the clinical spectrum of cystic fibrosis. We are not yet at the point where we can use prenatal diagnosis and DNA analysis to predict the severity of the disorder in a fetus, but a rough picture of the relationship between the presence of the common mutations and the expected severity of disease is slowly emerging.

The most exciting dividend of the cloning of the CF gene has been the tremendous impact it has had on driving creative approaches to new treatments. Thankfully, the health and life expectancy of patients with CF has, because of new antibiotics and better management of the risk for pneumonia, improved steadily for the last 30 years. Children born with CF today have an even chance to live into their late 30s, but advances have slowed, the disability can be severe, and the costs of care are high. For example, patients with end-stage cystic fibrosis are the people who most often undergo lung transplants or heart-lung transplants.

Because the disease is so common and because the life-shortening effects of CF arise as a consequence of recurrent damage to the lung, which is relatively accessible, CF is a focal point of gene therapy (see Chapter 20). Less than two years after the gene was cloned, scientists showed that they could use a virus to deliver an artificially constructed copy of the normal

CFTR gene into cultures of cells derived from a CF patient in which the gene was crippled and correct the defect in ion transport. Of course there is a huge difference between fixing cells in a test tube and treating humans, but hope is high. There are more than a dozen research projects under way in which the goal is to develop techniques to get normal CF genes into the lung tissue of affected children. Most focus on using relatively benign viruses that are known to penetrate human lung cells as "bio-missiles" to carry a payload of normal CFTR genes to the target organ. Some progress has been made. A decade from now there may well be an easy way to treat the disease, quite possibly with a "gene inhaler," a DNA spray that works much like the little pumps that persons with asthma use to deliver preset doses of medicine to their airways. But the discovery of the gene has also created sometimes vitriolic debate about genetic testing.

SCREENING FOR CF CARRIERS

In the fall of 1989, only weeks after the CF gene had been cloned, leaders of the American Society of Human Genetics publicly urged that the knowledge should not be used to screen the general population to identify carriers who, if they married a fellow carrier, would have a 1 in 4 risk in each pregnancy for conceiving a child with the disease. After two days of at times rancorous debate, a group of experts convened by the NIH in the late spring of 1990 to conduct a more thorough review of the issues reached the same conclusion. The scientific rationale was that it would not be helpful to screen the general population with a test that could identify less than 90% of those who actually carried a CF mutation. This is because such a test creates situations in which one partner definitely learns that he or she carries the CF allele while the true status of the other (who has tested negatively) remains unknown. Such couples have about a 1 in 1000 chance of having a child with CF. Some experts argued that couples in this situation might feel substantial anxiety in pregnancy. The experts also feared that if the test came into wide use too early most physicians in general practice would be unable to offer proper counseling about the implications of the result. Many on the NIH panel were also worried that carriers might be unfairly denied access to health or life insurance at reasonable rates.

At that NIH meeting I was among a minority that saw the matter differently. I argued that individuals have a right to ask questions about themselves, and that in an activity as crucial as childbearing there would be some couples who would be interested in learning about carrier testing for cystic fibrosis and who might well seek to be tested. I thought that the stance adopted by the majority was paternalistic, and that the patient's right to know should be inviolate. In addition, it seemed to me that the average family practitioner could readily be trained to provide the necessary information and support to patients who were thinking about whether or not to take the test.

During the 1990s there were many advances in the technical quality of the DNA-based test for cystic fibrosis, and it is available at reasonable price from many commercially and academically based labs. Currently, the most comprehensive test is offered by Genzyme Genetics, Inc., which in 1997 launched a test that could identify any 1 of 72 different mutations. Yet, as of 1999, CF testing is still not widely used. There were only about 75,000 carrier tests done that year in the United States (on a per capita basis, usage was somewhat higher in northern Europe). In general, most people who seek testing do so because they have a relative with the disorder or have symptoms that could be compatible with atypical cystic fibrosis, but the pattern of usage may soon broaden.

In April, 1997, another group of experts met at NIH to review progress in CF testing since the 1990 meeting and to reassess the place of carrier screening in medicine. I anticipated little change in their position, so I was quite surprised with their findings. In a cautiously worded statement, they supported offering the CF test to persons *regardless* of whether or not they had any family history of the disorder. This is a reversal of the position taken in 1990 and opens the doors to mass population screening. The 1997 position adopts an approach already in use in some prominent genetics clinics, such as that led by Dr. Arthur Beaudet at Baylor Medical Center in Houston, who advocates informing all patients, regardless of their family history, that a test is available to identify CF carriers.

One can imagine a number of different paths to develop routine population-based screening to identify CF carriers. Initially, the most efficient would probably be to offer the test with appropriate education and counseling to all young women who were planning soon to have families. Given the rapid shift to managed care, it would not be particularly difficult to incorporate a CF screening program into routine visits. If one assumes that 1 in 25 white women is a carrier, one can reasonably anticipate that about 1 in 28 white women would test positive for a mutation (the number is different because the test would not find about 10% of carriers). The next phase of testing would be to offer the CF test to the partners of the women who test positive. Again, about 1 in 28 of these men will be found to be a CF carrier.

Let us assume that a large managed care group adopted a CF carrier screening program and that 10,000 women in their 20s or 30s took the test in a year. Among them the test would find about 360 carriers. Among their 360 husbands, testing would identify about 13 carriers. Thus, by conducting 10,360 tests, the managed care group would identify 13 couples who are at 1 in 4 risk of having a child with CF. Without the program it is likely that among them, these 13 couples would have 3 or 4 first-born children with CF.

How does one decide if population-based CF screening is worthwhile? Assuming that it costs about $100 to test, educate, and counsel each person who decided to check her or his CF carrier status and to follow up with at-risk couples, the tab to identify four affected fetuses is about $1,000,000 or about $250,000 per case detected. The cost of finding a fetus with CF is considerably lower if one factors in savings from avoiding the birth of second affected children in some of the families, which often happens because of the delay in diagnosis that frequently occurs. Then the case finding cost might be closer to $100,000. This is still very expensive. Furthermore, either because of their religious beliefs or because they are impressed with the tremendous progress in caring for people with CF, about one-third of couples will not avoid at-risk pregnancies or abort an affected fetus.

One fact emerges from this back-of-the-envelope analysis. Until the unit cost of DNA testing drops by a factor of 2–5, it is unlikely that we will see a significant effort to screen millions of young Americans to identify CF carriers. The trump card that could refute this prediction is malpractice litigation. If even a single couple who gave birth to a child with CF successfully sued their physician for failing to alert them to the availability of the carrier test, CF testing would rapidly become widespread. During the late 1970s and early 1980s, lawsuits or fear of them accelerated the pace of prenatal screening for Down syndrome and for spina bifida (see Chapter 8).

The cost of DNA analysis may well drop by a factor of 5–10, probably in less than 5 years, and almost certainly in less than a decade, and it will continue to drop. To that virtual certainty add the fact that DNA testing will develop into an analytical system which is capable of asking thousands of questions about each blood sample regardless of what prompted the individual to seek information. Indeed, it might turn out (due to automation) to be easier and cheaper to ask thousands of questions about a sample than to ask only a few.

Assuming that the cost of DNA-based population screening for CF mutations falls to a few dollars per sample or less, the cost benefit analysis of finding the at-risk couple becomes compelling. It would then make good economic sense to identify those at risk, and to counsel them about the nature of the disease and their family planning options. This is the scenario that disturbs many individuals in the bioethics and disabilities communities. They fear that when costs of technology no longer constitute a buffer to widespread use, socioeconomic forces will drive consumers to use DNA testing and, for those who are positive, to avoid at-risk pregnancies or to electively terminate affected fetuses. Some see this as a sort of "neo-eugenics" (see Chapter 23) and warn of the return of a social policy in which the mantle of the state is taken up by the men who manage the flow of health care dollars.

In bioethics circles there is an apocryphal tale that officials at one HMO told the parents of a child with CF that should they decide to have another child, the HMO would only provide coverage of that child if the mother had prenatal testing for cystic fibrosis. In the early 1990s I was consulted on behalf of the couple that had reported this threat by their HMO. I was told that a HMO physician on one occasion did strongly suggest to a woman with a CF child that she should seek prenatal testing for future pregnancies, and that she perceived the suggestion to be a threat about loss of coverage. A single phone call to the HMO from the consulting geneticist caused the HMO to quickly distance itself from even a hint of coercive behavior.

Population-based screening to identify CF carriers raises thorny ethical questions about selective abortion. Is it right to provide DNA-based prenatal diagnosis to couples at risk so that they may have the option of pregnancy termination if the fetus is affected? Even though women have (since *Roe v. Wade* in 1973) a right to privacy that guarantees this option, a large fraction of the American population is opposed to it. In a 1997 poll of 819 adults conducted for *USA Today,* 49% said that they felt it was morally wrong to use prenatal testing to find and abort fetuses with cystic fibrosis. In the minds of many people the decision to terminate a pregnancy because one does not want to have a child is less offensive than is the decision to end a pregnancy because one is not satisfied with its anticipated condition.

Would some couples who learn that they are carrying a fetus destined to develop CF terminate the pregnancy? Yes. One survey of young adults who had no personal experience with CF indicated that about two-thirds would terminate. On the other hand, a survey of couples who had a child with CF found that less than one-quarter would electively abort an affected fetus. As the prospects for treating CF improve, interest in terminating such pregnancies will diminish.

Would young couples with no family history of CF seek carrier testing and, if they turn out to be carriers, avoid having a pregnancy that carries a 1 in 4 risk? They could circumvent the risk in a number of ways, including remaining childless, adopting, using a donor egg or donor semen, or using preimplantation genetic diagnosis (PGD). In PGD, CF testing is performed on a single cell taken from an 8-cell embryo conceived in a test tube with the couple's egg and sperm. This is possible because the laboratory can quickly make millions of copies of that cell's CFTR gene and scan it for mutations. If the embryo does not have CF (i.e., did not inherit two CF mutations), it is transferred to the woman by the same techniques used in standard treatments for infertility. The prospects for preimplantation diagnosis look excellent, but the cost is high, and at this stage the accuracy of the test results is sufficiently uncertain that physicians will only perform it if the pregnant woman agrees to undergo amniocentesis in order to obtain fetal cells to confirm that he or she does not have two disease alleles. Of course, this is a promise that the doctors cannot enforce.

Would mass population screening significantly decrease the number of infants with cystic fibrosis? A growing body of research suggests that once they learn about CF, the genetics of its inheritance, and the existence of a high-quality screening test, a sizable fraction of women want to be tested. However, we do not have a clue as to what impact mass population screening would have on the number of children born with the disorder. A few of the larger managed care organizations have started to offer CF screening routinely to young women, but it is uncertain how far or how fast this practice will spread.

Cystic fibrosis carrier testing may be among the first DNA-based tests for which population-based screening on a massive scale creates an early warning system for couples. However, the fact that genetic testing is so closely connected to selective abortion will continue to deeply trouble many people and cast a long shadow over molecular medicine.

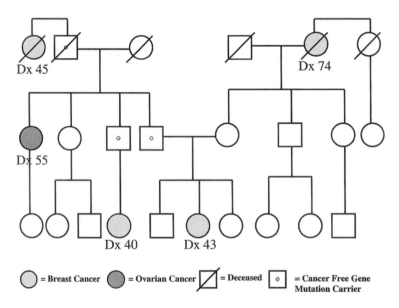

Dx 45

Dx 74

Dx 55

Dx 40

Dx 43

⬤ = Breast Cancer ⬤ = Ovarian Cancer ⬛ = Deceased ⬛ = Cancer Free Gene Mutation Carrier

A pedigree of a family burdened with hereditary breast and ovarian cancer.

Breast Cancer
The Burden of Knowing

GENES AND CANCER

Leafing through Molly Fitzpatrick's bulky family albums, one is struck immediately by the strong family resemblance. From grandmother to mother to daughters, from summer picnics to Christmas mornings, photo after photo has captured sparkling eyes, ivory skin, auburn hair, mischievous Irish smiles, and natural grace. The hundreds of photos, some almost a century old, record the rise of a family started by penniless immigrants whose descendants now enjoy the comforts of middle class life in the United States. But albums tend to record the good times, and candids cannot tell us what is going on inside. Only when Molly paused and touched a photo here and there and wistfully reminisced did I begin to understand the burden that her family has carried.

Over the last four generations, seven women on Molly's mother's side of the family have developed breast cancer and two have died of ovarian cancer. Of these nine women, only one has yet lived past 60 and most have died in their 50s. There are only two survivors, Molly's mother, Elizabeth, and her sister, Kathleen. After watching her mother and an older sister die, Kathleen, now 70, had her breasts surgically removed. At the time, more than 20 years ago, prophylactic mastectomy was an extremely unusual operation, and Kathleen had fought hard to convince the surgeon to operate. Elizabeth found her lump when it was small; she had her mastectomy 15 years ago.

Each year in the United States more than 150,000 women are diagnosed with breast cancer, and about 20,000 learn that they have ovarian cancer. During the year, 40,000 women will die with breast cancer and more than 10,000 will die from ovarian cancer. The lifetime risk of developing breast cancer is about 1 in 9, but it is important to remember that most cases occur in older women. In the general population, the risk of developing breast cancer by age 50 is "only" about 2%. The lifetime risk of developing ovarian cancer is about 1 in 65, and most cases arise in women over age 60. About 5–10% of the women who develop either of these cancers have a strong family history for one or both of them.

Physicians began writing about families with unusually high numbers of cancers more than a century ago. For decades such reports were treated as little more than medical curiosities, but as life expectancy soared (since 1900 the average life expectancy for white women born in the United States has climbed from 45 to 78, and it continues to rise) and we learned more about the diseases of old age, and our knowledge of pathology and genetics grew, the evidence that a significant fraction of breast cancer (and many other cancers) occurred in people who were born with a strong genetic predisposition mounted.

At the molecular level, every cancer is a genetic disease. We know this thanks to the work of many scientists over the last 25 years who have discovered and elucidated two classes of genes known as oncogenes and tumor suppressor genes. The several dozen oncogenes code for proteins that interact at key positions on the surfaces of cells and instruct them to grow and divide. They trigger a cascade of molecular messengers (a sort of biochemical pony express) that travel to the cell's nucleus and order certain genes to make the cell replicate. When they are normally functioning, oncogenes act to orchestrate the diverse aspects of human growth, the remodeling of injured tissue, and the replacement of dying cells. When oncogenes behave abnormally, the growth factors that they produce can cause massive overgrowth of cells, the essential feature of cancer. Signals to overproduce cells are, in most cases, not enough to cause cancer. To become the founder cell in a cancer, a cell must also overcome the forces of the tumor suppressor genes, proteins that normally act as restraints on growth.

Cancer arises when a single cell accumulates a sufficient number of mutations in several different genes that are each in some way responsible for regulating the cell cycle—the timing and pace of cell division. When these mutations significantly cripple the system that regulates that cycle, the cell may begin to replicate wildly. As further genetic changes occur, some progeny cells become so misprogrammed that they metastasize (break away and migrate to other parts of the body). This puts the individual at high risk of dying from metastatic cancer.

In most cases, the mutations in the oncogenes and the tumor suppressor genes (several dozen of which have now been identified) that lead to cancer were *not* present in the DNA with which the individual started life. They are somatic mutations that occur in cells in the various tissues over the course of one's life. Many cells in the body, especially those that are exposed to the outside world, such as those in lung and colon, must regularly divide to replace those that die off in the normal course of events. When they divide they pass on a copy of their DNA—including any newly acquired mutation—to the two daughter cells, which in due course may pass them on to their progeny. A few of these billions of daughter cells (there are on the order of 10 trillion cells in the human body) will in turn acquire additional mutations to other genes. Some of them can then pass on a bigger mutational load to their progeny. If a cell emerges that has inherited by chance just the right set of mutations, it has the potential to be the founder of a population of tumor cells. For some cancers we now know in exquisite detail the pathway to cancer. Among the most elegant work is that on the origin of colon cancer. During the early 1990s, a team led by Dr. Bert Vogelstein at Johns Hopkins University School of Medicine discovered which genes must mutate for colon cancer to arise.

No one understands exactly why most somatic mutations occur, but we have some good hypotheses. Certainly, as in lung cancer, there is overwhelming evidence that direct, repeated exposure to chemicals in cigarette smoke eventually cripples the key genes. There is also moderately impressive epidemiological evidence that a high-fat diet is a risk factor for cancer of the colon and, to a lesser extent, the breast. In fact, some of the nation's leading cancer epidemiologists think that smoking and a high-fat diet account for nearly two-thirds of all (non-skin) cancers in the United States. Given that percentage, none of the other major suspected causes actually constitutes a very large slice of the pie. For example, despite all the attention it has received, the low-level radiation to which we are all chronically exposed is thought to cause only about 2% of all cancer deaths.

Despite much concern about exposure to chemicals in the workplace, the epidemiological evidence suggests that it is a source of relatively few cancers, again probably only about 2% of the total. This is almost certainly much lower than was the case during the 19th and early 20th centuries. In fact, among the strongest evidence that chemical exposures in the workplace cause cancer is an observation made nearly two centuries ago. In the early 19th century, Sir Percival Potts, a British physician, noted that young men who worked as chimney sweeps often developed cancer of the scrotum. The cause was prolonged exposure of the skin to coal tars. Another well documented example of occupational cancer is the hugely increased rate of thyroid cancer among workers who were exposed to radium as they manufactured luminescent dials for watch faces (a practice now long abandoned). Today, thanks to better containment systems, the hundreds of dangerous chemicals that can be found in the workplaces of modern industrial society probably account for many fewer cancers than they did just 50 years ago. In poorer nations, which have much more dangerous workplaces, the risk of occupational cancer is probably much higher.

Most hereditary cancers, like those that have occurred in Molly Fitzpatrick's family, arise when several somatic mutations (changes in the DNA in cells in a particular organ) accumulate in the cells of an individual who was born with a germ-line mutation in a tumor suppressor gene. Because the germ-line change was present from the moment of conception, it is in the DNA of all cells in the body. In essence, people with germ-line mutations start life one big step closer to cancer than do those not so burdened. That is why one of the classic hallmarks of hereditary cancers is that they tend to strike at a younger age. Indeed, two of the cancers through the study of which the concept of tumor suppressor genes was in part elucidated, retinoblastoma (a tumor of the eye) and Wilm's tumor of the kidney, are childhood cancers.

BRCA1 AND BRCA2

In 1989 a team led by Mary Claire King, a scientist then at Berkeley, California, after painstakingly studying many families like the Fitzpatrick family, found convincing evidence that there is a gene on the long arm of chromosome 17 which, if mutated, greatly increases the lifetime risk for breast and ovarian cancer. The news set off a spirited race to find and clone the gene, a race that was won in 1994 by a team led by Mark Skolnick at Myriad Genetics, at the time a fledgling biotech company in Salt Lake City. Since then, the same research strategy has led to the discovery of a second gene predisposing to breast cancer, and in coming years it is likely that one or two more will be uncovered. Together, the two genes that have already been cloned, known as BRCA1 and BRCA2, account for much more than half of all hereditary breast and ovarian cancer. One of the curious facts about molecular genetics is that scientists now routinely find culprit genes before they have any idea about the gene's role in the life of a cell. We do not yet know much about what the BRCA1 and BRCA2 proteins do, although those secrets will soon be decoded.

Although the data are still being compiled, it appears clear that carriers of a harmful BRCA1 mutation have a 50–70% lifetime risk of developing breast cancer and a 30–50% risk of doing so by age 50. That is, their risk for early breast cancer is roughly 15 to 20 times higher than that of the average woman. Perhaps even more significant, but less discussed, is that their risk of ovarian cancer is also much higher than that of the general population. The lifetime risk for ovarian cancer in BRCA1 carriers is probably as high as 1 in 6, about 10–15 times higher than the background risk. It is crucial to inform women who do carry one of the harmful mutations about the risk of ovarian cancer. Indeed, a good case can be made that it is an even more important risk to discuss than is the risk of breast cancer. Most women are already highly aware that breast cancer is common, that there are simple methods (self-examination and periodic mammography) to detect most cases early, and that it is a highly treatable disease with steadily improving survival times. In general, women know far less about their risk of ovarian cancer. Furthermore, it is highly likely, although not yet absolutely proven, that in women known to carry a BRCA1 mutation, the removal of the ovaries greatly reduces the future risk of ovarian cancer. The surgery does not completely eliminate the risk because 5–10% of ovarian cancer arises from biologically similar cells that line the nearby pelvic wall which cannot be removed with the ovaries.

In 1996 scientists at Myriad Genetics launched the world's first laboratory test that reads the precise DNA sequence of a particular gene. Using highly automated techniques, the scientists at Myriad extract DNA from blood samples of at-risk individuals and examine about 16,500 DNA letters in the coding sequences of these two genes, searching for changes in even a single letter. This astounding new test (which on a single sample analyzes more DNA letters than all the laboratories in the world were analyzing in the mid-1970s) has created new hope and profound uncertainty for people like Molly Fitzpatrick. It also greatly exacerbated an often bitter policy debate about whether such a test was in the patients' best interests. Molly is among the first to face a dilemma that millions of us will experience in coming years. Should she find out if she carries a mutation that predisposes her to cancer?

Molly's mother, Elizabeth, developed breast cancer in 1982 at age 48.

Yet, Elizabeth considers herself lucky on four counts. First, her mother (Molly's grandmother) was diagnosed with breast cancer at 38 and died when she was 39, so Elizabeth thinks of herself as having eluded the disease for 11 years longer than did her mother. Second, because she regularly and meticulously examined her breasts, Elizabeth found her lump when it was small and she is doing well more than 15 years after her breast surgery. Third, she has not developed a second breast cancer or ovarian cancer, for both of which she is at high risk. She recently had her ovaries removed.

The fourth reason that Elizabeth considers herself lucky is that she knows she has a mutation in her BRCA1 gene. By becoming one of the first people to undergo the new DNA test, she converted a deeply held family suspicion to a fact. Elizabeth told me that she underwent the test for two reasons. For her the less important is that if she tested positive, it would provide a powerful rationale for having her ovaries removed. The more important is that she wanted to provide better information to her two daughters and her son about the chances that they could have inherited a gene that confers a significant cancer risk on them. Even though the absolute risk to her son remains low, men with BRCA1 mutations are at much higher risk for breast cancer than are other men. Early studies suggest that they may be at higher risk for prostate cancer as well. Because Elizabeth has a BRCA1 mutation there is a 1 in 2 chance that she passed it on to each of her children.

The fact that men are just as likely to carry BRCA1 and BRCA2 mutations and transmit them to their children as are women creates a clinically deceptive situation. In more than a few instances, women who turn out to have hereditary breast or ovarian cancer will be the first in their families to be afflicted. In such cases, the mutation may have passed silently through the male line. It could have arisen four generations ago in the sperm cell that started the life of a male fetus. In time the young man who was the product of that conception married and passed the mutation to both his sons. During the next generation perhaps each son, neither of whom developed cancer, had three children—two sons and a daughter. It would be quite plausible that only one of the two granddaughters inherited the mutation, which caused her to develop breast cancer when she was, say, 52. When that woman recounted her family history for her doctor, she would disclose nothing that would suggest that she was born with a hereditary predisposition to breast cancer.

Scientists were able to find the BRCA1 and BRCA2 genes by studying unusually large families in which many people had been diagnosed with breast cancer. These represent the tip of an iceberg of unknown dimensions. As we learn more, it will probably turn out that the genetically based risk for breast and other cancers is somewhat higher than we now think, but that the risk is spread over many different genes, some of which contribute only a small component to the overall predisposition. The role of still poorly understood environmental factors in breast cancer is and always will be high.

Her mother's decision to be among the first people to undergo a DNA test for cancer risk has had a huge impact on Molly. The positive result means that she has a 1 in 2 chance of having inherited a gene that (almost certainly) caused her mother's breast cancer. She could find out at any time if at conception she won or lost that particular toss of the genetic coin. Molly has agonized over whether or not to be tested. On the one hand, the possibility of learning that she won the toss and did not inherit the chromosome with BRCA1 mutation has tremendous allure. If she is not a carrier, many of the fears that trouble her would vanish. She would no longer be terrified about dying young of ovarian cancer as did one of her aunts, she would stop wondering if she should have her healthy breasts removed to reduce her risk of cancer, and she would stop worrying about whether she will transmit a predisposing cancer gene to her children when she has them.

What if she took the test and the results showed that she did inherit the mutation from her mother? All those worries that she is usually able to keep at bay may mushroom. Old questions may take on new urgency. Should she consider mastectomy? There is some fairly persuasive evidence that it saves the lives of BRCA1 or 2 carriers. In 1996 Lynn Hartmann, a researcher at the Mayo Clinic, published her findings from a study of women who, in addition to having a breast removed to treat an existing cancer, also had the other removed to avoid a future cancer. In comparing them to similar women who did not have the second surgery, Dr. Hartmann found fewer deaths from breast cancer in the group that had both breasts removed. This was the first evidence that prophylactic surgery saves lives, but as it was a retrospective study, it left some scientists unconvinced.

Imagine the difficulty of deciding whether or not to undergo prophylactic surgery. Currently, few women request this option and few surgeons recommend it, largely because they believe that by carefully following women at high risk they can find a breast cancer so early that it will be highly curable. On the other hand, a growing number of medical centers are offering surgery to remove the ovaries of women who carry a deleterious BRCA1 mutation because it is not yet possible to monitor such women to catch ovarian cancer early enough to cure it. Molly, who is 30 and soon to marry, hopes to have two children over the next four years. Assuming no major breakthroughs in chemoprevention that would allow her to avoid surgery, she currently plans to have her ovaries removed before she is 40.

The hunt for genes that may predispose to breast and ovarian cancer has explained an old and puzzling observation. Ashkenazi Jewish women (about 90% of the Jews in the United States are of Ashkenazi descent) have an above average risk for breast cancer. For years this increased risk was attributed to unknown factors in diet or lifestyle. We now know that the reason is largely genetic. Two mutations in the BRCA1 and one mutation in the BRCA2 gene are unusually common among the Ashkenazim. Overall, an Ashkenazi Jew has about a 1 in 40 chance of being born with one of these mutations. The high (from a geneticist's point of view) prevalence of these mutations in this population is probably due to a combination of three factors. First, because the Ashkenazim constitute a relatively small population group whose members have through most of history tended to marry other members (a pattern that geneticists call endogamy), it is likely that some mutations that appeared by chance were propagated. Second, since these mutations, which usually cause illness well after the individual has had children, do not have any negative effect on reproductive fitness (the likelihood that each person has of passing on his or her genes), they will not tend to drain out of the gene pool. Third, because the three mutations are so common, there is even a chance that they may confer some (as yet completely unknown) benefit during the first four decades of life.

Because only three mutations explain most of the hereditary risk for breast cancer in the Ashkenazi population, it has been technically easy to launch a low-cost DNA-based test for Ashkenazi Jewish women. So far, however, relatively few people have made use of it. Because there is no simple medical intervention, such as taking a medication with few side effects, that will reduce the risk faced by carriers, and because few women want to undergo bilateral mastectomy, many Jewish women are debating whether learning that they do carry one of the three mutations will be helpful. Since breast cancer is a risk faced by all women and since established surveillance techniques will, if carefully adhered to, catch most breast cancers at a relatively early stage, why not just take monitoring seriously? This might well work, but it does not solve the problem that being a BRCA1 carrier also confers a significantly increased risk for ovarian cancer for which there is not a good surveillance system. Ovarian cancer is an insidious disease that in the vast majority of cases is only diagnosed in an advanced stage, and is usually fatal in less than five years. If, like Molly, one has a significant family history of breast cancer and one can learn whether her risk for that disorder and for ovarian cancer is or is not elevated, the information is potentially of great value.

Another major worry articulated by many women with a family history of breast cancer (and a major concern of the National Action Plan on Breast Cancer) is that when a healthy individual learns that she carries a mutation that predisposes to breast cancer, she and her relatives will have trouble obtaining health and life insurance. Worries about genetic discrimination are understandable. Genetic ideas were misused in the United States for much of the first half of this century, leading to the involuntary sterilization of about 60,000 persons with mental retardation and to national immigration quotas keyed to notions of relative racial superiority (see Chapter 24). Socioeconomic discrimination on the basis of race and gender (both genetically determined) remains widespread today. People with cancer have traditionally had great difficulty in finding affordable health insurance and life insurance.

What evidence is there that healthy people in the United States are being denied access to health insurance or charged higher premiums because of a genetic test that indicates an increased risk of a serious disease or of bearing a child with a genetic disorder? In a word—almost none. In 1992 and again in 1996, a Boston-based group of scientists published papers based on surveys of families at risk for various genetic disorders which purported to show that a significant fraction had been denied health insurance. But, because the authors did not contact the insurers to study their side of the story, it is impossible to verify the claims of the consumers. The 1992 paper generated a lot of media attention and public interest. Since then I have frequently sought firm evidence of genetic discrimination in health insurance underwriting and have repeatedly come up empty. The insurance executives with whom I have spoken deny any such practice. They point out that (1) almost all health insurance is group-written so that the health of any single person or family is largely irrelevant, (2) the cost of asking people to take genetic tests as part of their application is far too expensive (for now) to justify its use, and (3) it would be legal and political suicide to be caught engaging in genetic discrimination.

The era of DNA-based testing to find out whether one is at special risk for breast or other cancers is just beginning. Thus far, there is little evidence to suggest that persons who test positive for BRCA1 or BRCA2 mutations or for other genes that cause risk for other cancers have had trouble obtaining health care coverage. Indeed, in the only study conducted of a large cohort of women who had undergone the BRCA1/2 test offered by Myriad, the vast majority indicated that they were glad that they had taken the test and had not suffered any economic discrimination as a consequence of so doing. This work was done by researchers who were not associated with the company. Most Americans now obtain health care coverage either through federally funded programs such as Medicare or Medicaid or through employer-based group health insurance. On August 1, 1997, the federal Health Insurance Portability and Accountability Act (HIPAA) went into effect. That law expressly forbids group health insurers from defining predictive genetic information as a "preexisting condition" (which would give the insurers a right to deny coverage for a particular condition for up to 18 months). In the summer of 1997, President Clinton put genetic discrimination on the national agenda when he came out strongly in favor of laws to prevent it. Several bills that would, if enacted, broaden the protections provided by HIPAA are now before the Congress. The protections offered by HIPAA as well as a steadily growing number of state laws (at the close of 1999 there were at least 30) that forbid health insurers from using predictive genetic information in underwriting, suggest to me that our society has done a pretty good job of heading off a possible new form of discrimination.

Nevertheless, a new problem has emerged. Advocacy groups, bioethicists, lawyers, and some leading scientists have been so effective in pointing out the potential threat of genetic discrimination that many people who are members of families with a history suggesting that they might carry mutations like BRCA1 or BRCA2 are telling their physicians that fear of discrimination is the main reason they will not take the predictive test. Even though there is little direct evidence to justify that fear, if the perception cannot be changed, it defines reality. Hopefully, over the next two or three years people will be gradually reassured by the enactment of laws to ameliorate this fear. If not, we may face the tragic situation of people refusing to undergo DNA tests that could eventually lead to life-saving strategies in the management of cancer risk.

*(Left to right) Alois Alzheimer, Emil Kraepelin, Robert Gaupp, and Franz Nissl.
(Reprinted from Pollen 1993 [photo from Dr. Paul Hoff, Psychiatric Hospital of the Ludwig–
Maximilians–University, Munich].)*

Alzheimer Disease
Are You at High Risk?

In November 1994, Ronald Reagan announced that he was suffering from Alzheimer disease (in using eponyms to designated diseases, the powers that be recommend avoiding the possessive). It was the last public statement he was to make. Since then, those around him have been chary of saying much, but it appears that the former President has declined precipitously. Reagan no longer greets admirers with his trademark, slightly startled way. He cannot carry on a conversation and he recognizes only a handful of people. His memory is now severely impaired. Nancy Reagan does not permit journalists or, for that matter, any but a few close friends to visit him. The gipper is housebound in their BelAir mansion, and his beloved ranch in the Santa Ynez Mountains was sold years ago. Unfortunately, things will get even worse.

Alzheimer disease begins with an insidious loss of memory and other cognitive skills and ends with the patient bedridden, incontinent, unable to speak, barely aware of family and friends, and totally dependent in all aspects of daily life. The decline is slow but inexorable, and extremely painful to loved ones. The only possible blessing is that before the end, the individual loses awareness of the depths of his or her illness.

My first clinical contact with a person with advanced Alzheimer disease occurred when I was a medical student. I was assisting an intern in the workup of a man with a high fever. As we prepared to do a spinal tap (a standard part of the workup of dementia of unknown cause), the intern repeatedly told the man not to move. The patient repeatedly said, "O.K." Just after the young doctor had successfully inserted the 4-inch needle into the spinal canal, the patient stood up and ran out of the room, naked, the needle bobbing in his back. We gave chase and, with the help of two nurses, physically restrained him for his own safety. He was not being obstreperous; he simply could not remember even for seconds the simple instructions that he had been given.

Ronald Reagan is the most famous of the approximately eight million Americans who suffer with Alzheimer disease. Less well known is the fact that his brother, Neil, had advanced Alzheimer disease when he died in December of 1996, and (although awareness of the disorder was much lower back then so we cannot be sure of the diagnosis) that his mother, Nellie, probably also was affected. The Reagan family's troubles emphasize an important point. This most common disorder of old age is often familial and arises in many cases due to a genetically determined predisposition.

HISTORY

The first person to be diagnosed with Alzheimer disease was a middle-aged German woman who had begun to suffer severe memory loss at the age of 51. She was followed for several years by Dr. Alois Alzheimer, a 43-year-old research neurologist and psychiatrist then working in Munich. When the woman died in 1906, Alzheimer obtained permission from her family to perform an autopsy and to study her brain tissue. A year later he published a short paper describing the severe dementia from which she had suffered (before death she had lost her ability to speak) and hypothesized that her illness was caused by the presence of unusual-looking material that he had seen scattered through her brain cells. Alzheimer was an astute observer. The original case report includes a clear description of the plaques and fibrillary tangles that are today still the microscopic hallmark of the disorder. Because there are no biochemical markers for the disease and because dementia has many causes, an unequivocal diagnosis still can only be made by autopsy.

In 1907 the pathological findings in senile dementia, the dementia seen in otherwise healthy old and very old people, had already been characterized. Physicians thought that most cases arose due to vascular disease (hardening of the arteries). They also were aware that dementia occasionally arose for no discernible reason in much younger adults in their 40s and 50s, but until the research by Alzheimer, no one had formally studied their brain tissue after death.

Alzheimer's work led Emil Kraepelin, at the time the world's leading proponent of a biological view of psychiatry (he was one of first physicians to study schizophrenia), and one of Alzheimer's mentors, to conclude in 1910 that his student had described a unique "presenile" form of dementia. It was Kraepelin who suggested that this clinical entity should be named in honor of his student, a recommendation powerful enough to ensure that the honor was bestowed. Unfortunately, Dr. Alzheimer developed rheumatic heart disease, and died in 1915 at the age of 51, just as World War I was curtailing most medical research. After the war no one took up the work that he had begun, and two generations slipped away before the disease he characterized again received serious scientific attention. Writing about Alzheimer disease in 1948, R.D. Newton, a British neurologist who was instrumental in stimulating renewed interest in presenile dementia, lamented that in four decades there has been "little advance towards a solution of the problem."

The hiatus is understandable. During the period from 1910 to 1950, physicians faced many other more urgent clinical problems. In the early years of this century, the median life expectancy in the United States and western Europe was still less than 50. Tuberculosis (known as the white plague) was the number one cause of death, childhood mortality was immense, and childbirth was a life-threatening experience. Streptococcal pneumonia killed hundreds of thousands of people each year. The diseases and disabilities of the elderly had a low priority in American and European medicine. Geriatrics was not yet dreamed of as a medical specialty. Because few people studied the elderly, most of what we now know to be diseases of old age were viewed as the normal, inevitable, if unfortunate, consequences of aging.

The rise of Alzheimer disease as a public health problem of gargantuan proportions is a direct result of the tremendous medical advances of the 20th century. By the third decade of the 20th century, the public health was rapidly improving. Beginning in the late 1930s, anti-tuberculosis drugs and public health measures were combined to rapidly reduce deaths from that disorder. With the development of antibiotics in the 1940s and 1950s, and vaccines in the 1950s, childhood mortality fell precipitously. Over a stretch of only 50 years the median life expectancy for Americans rose by 20 years, an advance unequaled in all history and likely never to be repeated. Suddenly, many more people in the United States and Europe were living into old age. During the 1950s and 1960s, cancer, heart disease, and stroke steadily assumed much more importance on the national health agenda. Even then, dementia was largely characterized as a consequence of aging rather than as a set of distinct disorders that could be diagnosed and treated.

Although little was yet understood about it, two interesting aspects of Alzheimer disease were recognized in mid-century. A few reports of families in which many persons suffered from early onset of dementia were published in the German and English medical literature in the 1930s and 1940s, for the first time raising the issue of genetic predisposition. Furthermore, by comparing the prevalence of early disease among identical and nonidentical twins, Franz Kallman, a pioneer in the search for genetic factors in complex disorders, found strong evidence of a genetic risk. Another important advance was that neuropathologists could find no differences when they compared the lesions in the brain tissue of relatively young persons who had died with the disorder with the specimens taken from old persons. Contrary to decades of thinking that the two groups must suffer from different disorders, the evidence suggested that they both suffered from forms of the same disease.

The first crude (and unsuccessful) attempt at using linkage studies to pursue the hypothesis that Alzheimer disease may be heavily influenced by genetic predisposition was made in the late 1950s when a researcher made a long-shot bet that risk for the disorder might in some way be correlated with having a particular blood type. In 1960 there was a second unsuccessful, but poignant, effort to determine whether predisposition to the disease could be linked with having a particular blood marker among those in the MNS red cell antigen system. It was conducted by a physician who was a member of a rare family in which risk for the disorder seemed to be due to the effect of a dominant gene.

An important turning point came in 1969 when the Ciba Foundation held an international symposium to assess the status of knowledge about Alzheimer disease. This helped to reconceptualize it as a distinct disorder, not a phenomenon of aging. During the 1970s a neurologist, Robert Katzman, who had begun studying the biochemistry of Alzheimer disease in the 1960s, emerged as a leading proponent for funding in this field. His efforts were timely. In 1974 the federal government created the National Institute of Aging (NIA). The 1970s saw a great increase in public interest in Alzheimer disease. The main reason was that the more carefully physicians and epidemiologists investigated the disorder, the larger became their estimates of its prevalence.

Interest in the use of molecular markers to find genes that predisposed to Alzheimer disease received great impetus from the stunning success in 1983 of a team led by Jim Gusella, a molecular biologist at the Massachusetts General Hospital, in mapping the gene for Huntington disease, another neurodegenerative disease, to a tiny region on the tip of chromosome 4. At the time, efforts to map disease genes by showing they must be located near established DNA markers seemed quixotic. Given the vast size of the human genome, no one thought that Gusella and his colleagues had a chance. What had seemed impossible suddenly appeared possible. When they quickly succeeded, the whole scientific world took notice. A physician-scientist named Peter St. George-Hyslop quickly took up the challenge to repeat Gusella's feat with Alzheimer disease. He arranged with Gusella to do postdoctoral studies in his laboratory so he could learn to do linkage studies. Long intrigued by the observation that virtually all persons with Down syndrome eventually develop Alzheimer disease, and knowing that persons with Down syndrome have an extra chromosome 21, St. George-Hyslop became one of a handful of people to focus his efforts on that relatively small region of the genome.

During 1985 and 1986, a team that included St. George Hyslop, a young molecular biologist named Rudi Tanzi who was constructing a critically important genetic map of chromosome 21 (and, thus, providing the markers without which a linkage study could not be attempted), and Daniel Pollen, a neurologist at the University of Massachusetts Medical School who had been gathering data and blood samples from a huge Alzheimer disease family of Russian extraction, worked furiously to investigate the hypothesis that at least one gene that predisposes to Alzheimer disease must be hidden on that chromosome. Finishing in a dead heat with three other research groups, they were able to publish strong statistical evidence for the existence of such a gene on 21 in the winter of 1987. At almost the same time, several other groups showed that a gene for a protein called "amyloid precursor protein (APP)" also mapped to chromosome 21. Since excess amyloid is the material that makes up the extracellular plaques that Dr. Alzheimer discovered in the brain cells of his deceased patients when he studied them under the microscope, the possibility that a defect in the APP gene causes at least some cases of the disease was irresistible.

In February 1991, a team led by John Hardy, a British geneticist, became the first to show that Alzheimer disease could be caused by a mutation in a gene. He found that in one of the families in which Alzheimer disease behaved as though it were an autosomal dominant genetic disorder, there was a mutation at a certain spot in the APP gene that was always present in the DNA of affected persons and always absent in unaffected relatives. One measure of the importance of this discovery, the first irrefutable proof of why some people develop Alzheimer disease, is that their announcement became the most frequently cited paper in the entire scientific literature for that year! Only later did it become clear that this mutation accounted for an exceedingly small proportion of all cases of Alzheimer disease—about 1 in 200. Hardy's team had been studying a fascinating, but exceedingly atypical, family.

As the set of reference markers for the map of the human genome grew ever more dense during the early 1990s, the prospects of finding rare genes that predisposed to Alzheimer disease greatly improved. From 1990 to 1994, the gene mappers made tantalizing progress in finding genes that caused very rare forms of the disorder. In October 1992, a team led by Gerard Schellenberg of the University of Washington reported that in seven of nine Alzheimer families in which the disorder had very early onset, they had found strong evidence for a gene on the long arm of chromosome 14. About two years later, Schellenberg and other researchers, studying a group of families known as the Volga German kindreds (who had emigrated from the Hesse region to Russia in the 1760s and extensively intermarried), found strong evidence of linkage to another gene on chromosome 1. By 1995 the predisposing genes on both chromosomes had been cloned, and study of their proteins was under way. In both cases, however, the families that had provided the clinical information and DNA samples were, like those studied by Hardy, atypical. As recently as 1993, no one had a guess as to the location of the big gene, the one that researchers would be able to associate with increased risk for Alzheimer disease, in millions of people.

APOE4

As so often happens in science, serendipity played a huge part in the next discovery. In the early 1990s, Warren Strittmayer, a biochemist who was primarily interested in cholesterol metabolism and aging, joined a team at Duke University School of Medicine led by Allen Roses, a prominent neurologist who had for some years been especially interested in the genetics of late-onset Alzheimer disease. In 1991 a team led by Roses had reported that linkage studies with late-onset families suggested there was an influential predisposing gene on chromosome 19, but the observation did not at first generate much interest in the then relatively small Alzheimer research community, most of whose members were focusing on working out the genetics in the much less common, early-onset families. When Strittmayer heard about the statistical evidence that there was a predisposing gene somewhere on a region of chromosome 19, he realized that a gene that he had been studying that coded for a protein known as apolipoprotein E (apoE) that was involved in cholesterol transport was in the same region. Since defects in cholesterol metabolism can cause coronary artery disease and stroke, Strittmayer and Roses reasoned that a particular variant in the apoE gene might affect the brain in a way that predisposed those who had it to Alzheimer disease.

Roses quickly refined the hypothesis. In essence, he asked whether one could correlate the presence of Alzheimer disease with one of the three subtypes of apoE4 (called E2, E3, and E4) that are found among humans. The results were dramatic. He found that those individuals born with at least one copy of apoE4 (about 15% of all alleles in the population) were twice as likely to develop Alzheimer disease as were those who were born with apoE2 (about 7% of all alleles) and/or apoE3 (about 78% of all alleles). Roses' early research suggested that the 2% of the population born with two copies of apoE4 (15% times 15% is about 2%) were 9 times more likely than those with two copies of apoE3 (about 62% of the population) to develop Alzheimer disease. Furthermore, those with apoE4 appeared on average to develop symptoms at an earlier age than did those with apoE3 who developed the disorder. Roses and his team had cut the Gordian knot that everyone else had been trying to unravel. They had shown that there was a common gene variant that could explain a major portion (at least half, and perhaps as much as three-quarters) of the risk for developing Alzheimer disease in millions of people.

Once disbelief gave way to excitement, scientists rushed to test Roses' findings. In a little over two years, more than 90 studies in populations around the world proved that the association was real. As knowledge grew, it became apparent that the three slightly different proteins made by the three subtypes of the apoE gene strongly influenced the *rate* at which the age-related risk for Alzheimer disease grew. As Roses put it in one article, the data indicated that if humans all lived to be 140 they would all develop Alzheimer disease, regardless of their apoE status. However, the particular dyad of apoE alleles with which one was born strongly affected how *early* the degenerative disorder might appear. apoE4, which apparently is more efficient than is E2 or E3 in assisting the transport of the protein called amyloid into brain cells, moves one more quickly to the threshold of cellular injury at which Alzheimer disease will develop. On the other hand, if one is lucky enough to have been born with two copies of apoE2 (less than 1% of us) it appears that one's risk for early Alzheimer disease is unusually low. Persons who have two copies of apoE4 typically develop Alzheimer disease 20 years earlier than do individuals who have some combination of apoE2 and apoE3.

The Testing Dilemma

Almost overnight, knowledge about the relationship between apoE status and risk for Alzheimer disease emerged as a major ethical issue in medicine. Unlike many developments in genetics, this one involved information that could be fairly easily obtained from a well-established, widely available, low-cost test (apoE testing had been used by cardiologists studying lipid metabolism in patients with high cholesterol levels). It would be difficult, if not impossible, to curtail its use. Yet, apoE testing to assess risk for Alzheimer disease only permits one to offer a general statistical likelihood. apoE status is most definitely not tantamount to a diagnosis, nor does it approach certainty as a predictor that any particular individual will develop Alzheimer disease. Furthermore, even for a person in the highest risk group (e.g., with two copies of the E4 gene) one cannot yet responsibly hazard a guess about likely age of onset of the disease, should it develop at all.

Roses cautioned the medical community that there was no basis upon which apoE testing could be justified as a test for population-based screening or even for targeted screening in families with worrisome histories. He and others argued that its best use was to support the diagnosis in persons with discernible signs of dementia. A second possible use is to try to identify which persons with Alzheimer disease should be tried on a medicine called Tacrine, which is of questionable value but appears to slow the rate of memory loss depending on one's apoE status. During 1995 a number of papers were published that substantially agreed with his views, and in 2000 the official word on apoE4 testing is still that it should not be used to predict risk. Nevertheless, hundreds of thousands of Americans, especially those with a parent who has been diagnosed with Alzheimer disease, have asked their doctors about testing, and many thousands have been tested. The question of access to apoE testing has become a classic test case of the doctor as gatekeeper, of paternalism in medicine.

I first faced this gatekeeping test in the fall of 1995. A friend who is a successful, healthy, middle-aged accountant was struggling to care for his father during his last months with Alzheimer disease. My friend had read about apoE testing in the *Wall Street Journal* (which has a long history of excellent reporting on advances in medicine). When he asked his father's family practitioner about being tested, the physician had dismissed the suggestion as unlikely to yield helpful information. My friend thought differently. He wanted first to have his father tested. If his father had one or two copies of apoE4, then he wanted to be tested. He reasoned that if his father had two copies of E4 and he had only one, then his risk for developing the disorder would be lower. Similarly, if his father had one copy of E4 (as do 15% of all Americans and 50% of those with Alzheimer disease) and he had none, he could also presume that his risk was less than his father's had been.

In both instances he was correct, but he was ignoring a lot of other important facts. For example, about one-half of patients with late-onset Alzheimer disease do not carry any apoE4 alleles. Furthermore, more than half the people with one copy of apoE4 never develop Alzheimer disease (probably in many instances because they do not live long enough). Even if my friend turned out to have two copies of apoE4, the worst possible, but most predictive, scenario, it was by no means certain that he would develop the disease, nor, if he was destined to, could one predict at what age. Yet, if he did turn out to be homozygous for apoE4 might he (and perhaps others in his family) begin to think of Alzheimer disease as his destiny? What impact might that have on his mood, his self-esteem, his performance in the workplace? Would undesired test results change his low-level anxiety and reactive depression into a more serious depression?

Other issues also surfaced. Depending on how I sent the blood sample off for testing and how he decided to pay for the test, the results might or might not wind up in my friend's medical record. What implications would flow from that? If it turned out that he has two copies of apoE4 and he told this to his wife and children, how would they react? Would a report that he had one copy of apoE4 suggest to a lay person (such as a clerk at an insurance company) that he might no longer fall into the normal pool of insurable risks? Why be tested, such a person might reason, unless the results mean something? Would this compromise his ability to get disability or long-term-care insurance at standard rates? Was it proper for me, as his physician, to create a shadow chart, one that sequestered medical information from other doctors? To absolutely protect his privacy, should I (if I agreed to order it) test his sample under a pseudonym, and pay for it with a check drawn on an office account? Is there a place for anonymous testing in genetic medicine? Was I becoming paranoid?

I urged my friend not to have his father tested or to be tested himself, and he agreed, albeit reluctantly. I think he acquiesced because he did not want to raise our disagreement to the level of a confrontation. What right did I have to deny him information about himself? After all, he was not asking me to prescribe him a medicine that he did not need and that might harm him. An accountant who was far more adept at considering statistics than am I, he was asking me to help him obtain information that in the face of a positive family history might assist him in recalculating his risk for developing Alzheimer disease. What is wrong with that? The standard answer has been that most people are not sophisticated enough to assess the information they seek. Geneticists have feared that without expert counseling (and all too often, even with it), patients will misapply the information or that others will misapply it in making judgments about them. This argument is of little persuasive power to well-informed persons, nor should it be. We clearly have much to decide about how to use and how not to use predictive genetic testing.

The overwhelming opposition among human geneticists, neurologists, bioethicists, and others to using apoE testing to assess risk for Alzheimer disease that was manifest in the mid-1990s is already beginning to elicit counter-arguments. Some physicians and consumers are asserting that any test that can suggest whether or not one is at increased risk for developing a severe, late-onset disorder and, if so, on average nearly 20 years earlier than those with the standard background risk, is actually of immense value. The result can reshape a person's plans for career, retirement, estate planning, and a host of other major life issues. Between 2000 and 2003, Dr. Robert Green of Boston University School of Medicine and his colleagues at several other institutions will be studying a large cohort of persons with a family history of Alzheimer disease to determine how they respond to learning their apoE status. If the research shows that information is of value to the individuals, use of apoE testing may become more common.

Assuming that physicians decided to recommend careful use of apoE testing, they and their patients would run into another problem. A single company controls the intellectual property governing use of the test for this purpose. During the late 1990s, it was reluctant to permit other labs to license the right to test at typical industry rates. This situation (which is true for a growing number of genetic tests) constitutes an effective monopoly in which one lab sets the price of the test and forces all samples to flow to it. The implications for patients are obvious; they pay more. This is just one of a burgeoning number of intellectual property issues that will cause significant problems over the coming years and that cry for novel resolution.

Does Ronald Reagan have at least one copy of apoE4? Given his family history, I would guess that he does. Does it matter? Not yet. Someday, when we are able to fashion therapies in part keyed to genotype, it may matter a lot, but that day is still far off.

W. French Anderson, M.D.

Gene Therapy
The Dream and the Reality

REMEMBERING DAVID

Severe combined immune deficiency (SCID) is a term used to describe several exceedingly rare, but cruel, genetic diseases that mimic AIDS. Affected children are normal at birth, but within months, as the protective antibodies that their mothers provided to them disappear, they are plagued by repeated bouts of unusual infections. They fail to grow well and are often hospitalized for pneumonia. Because of a single small genetic error, their immune systems cannot properly make the T cells and B cells that normally patrol the human perimeter and quickly kill foreign bacteria, fungi, and viruses that are incessantly trying to invade.

Despite its rarity, SCID became a famous disease for a time in the 1970s thanks to a boy named David who lived near Houston, Texas. David was born with an X-linked version of SCID. When it became clear that David was severely affected and would likely die within months or a few years, his parents made the dramatic decision to raise him in a sterile environment. Working with physicians and scientists at Baylor Medical Center, they created a special home for him, a large clear plastic bubble, which was, in effect, an immunological space station. As David grew, the rooms were enlarged, a special van was outfitted to transport him back and forth from his suburban home to the medical center, and a sterile room was set up for him at the hospital. David lived like this for 12 years. After he entered his sterile world, a world almost devoid of microorganisms, a world in which his food was irradiated, David took on a life that no other human had ever experienced. He could see and talk with his family, but he could not sit down to dinner with them. He could be lovingly touched across the flexible plastic walls, but he could never feel the warmth of his mother's kiss. Friends could engage in parallel play with him at the edge of his life-giving prison, but he could never swing side by side with them at a playground.

In taking David on this extraordinary journey, his parents were fighting to buy time. They were hoping that medical research would come up with a treatment that would free him. Years passed and David became famous. Science journalists loved to report about him, and he became the subject of a made-for-television movie. But there was no major advance in understanding his genetic disorder, and the dream of finding a way to repair his faulty gene remained distant. David's parents knew that the bubble solution could only work for so long. They guessed correctly that as he approached adolescence and realized his situation, David would want to try the only possible cure, bone marrow transplant surgery, even though at the time it carried a substantial risk of death. Since David's disease resulted from defective white cells that are made in the bone marrow, doctors could try, using drugs and radiation, to destroy his marrow entirely, and then reconstitute it with donor cells. If it worked, David could live a normal life. If it failed and David developed a massive postoperative infection that could not be successfully treated, he would die.

When he was 12, David told his parents and his doctors that he wanted to take the risk. Unfortunately, heroic efforts to find a bone marrow donor who would be a perfect immunological match were unsuccessful. There was one other choice. Scientists at the Dana Farber Cancer Institute in Boston had developed a technique to help imperfectly matched bone marrow donations work. David desperately wanted to go forward. In October of 1983, the clinical researchers took bone marrow donated by his sister and prepared it for transplant into David. He was operated on the next day, and quickly put back in a sterile chamber while the doctors waited to see if his body would make a new immune system. At first things seemed to be going well, but in December David developed a bad infection and became very ill. As he got worse and was moved to the hospital, the doctors realized that he was suffering from an incurable infection with Epstein-Barr virus. The virus, to which virtually all of us have been exposed, usually without incident, had apparently been a hidden invader in the donor bone marrow cells. With nothing to stop it, the bug caused a runaway cancer of David's B cells, a kind of lymphoma. His death in January of 1984 was among the most emotionally wrenching losses in medicine, so invested had been the scores of people who had cared for him. Just seven years later, doctors opened the doors to the world of gene therapy by treating a little girl with the same disease.

GENE THERAPY: HISTORY AND CONTROVERSY

A few passionately committed individuals, especially Dr. French Anderson, who at the time was directing a laboratory at the National Institutes of Health, started pursuing gene therapy long before there was any reason to hope that it might work. Their collective idea was this: Imagine a serious, untreatable disorder that arises because a defect in a single gene prevents a particular organ or cell type from working (diseases arising in the bone marrow were the first obvious candidates). Perhaps it would be possible to isolate a normal version of the disease gene, amplify it many fold, and attach millions of copies of it to vectors (submicroscopic vehicles such as viruses that would carry it into other cells). One could remove the target cells from the patient, stimulate them to grow in culture, put the vectors carrying the normal version of the gene in the culture, and return them to the individual. If the vectors entered those cells and delivered the normal gene to the nucleus, perhaps some of them would insinuate themselves into the patient's genome and begin making a normal protein. If such cells were then returned to the patient and if the donor genes kept working, maybe the disease in question could be ameliorated or cured. If such an effort worked even once it would usher in a new era of molecular medicine.

The term "gene therapy" includes two vastly different concepts. The notion of using DNA vectors to treat disease in a human is known as somatic (or body) cell gene therapy. From the start the only major ethical concerns generated by this dream have focused on safety. How could one possibly calculate the risk to the patient of deliberately infecting his cells with a foreign gene that might incorporate anywhere in his genome? Did this create a risk of cancer? Of course, for desperately ill patients for whom all other therapeutic options have been exhausted, the idea of undergoing an experimental therapy and taking novel risks becomes more acceptable, both to them and to society.

The second, far more controversial, idea is germ-line gene therapy, which has never been attempted in humans. This would involve the genetic manipulation of a single egg or sperm cell prior to conception or, more likely, the genetic manipulation of a 4- to 8-cell embryo created in a test tube. A gene or genes added at this early stage of development would become part of the cells that mature to make all of the cell lines that organize during embryogenesis, including the future individual's sperm or eggs. That is, the work of the genetic engineers could be passed on through the patient's descendants, spreading ever so gradually as a tiny ripple in the human gene pool.

Although this is still not technically possible in humans, scientists and bioethicists have been debating the morality of germ-line genetic therapy for 30 years. Some have worried about creating unpredictable biological risks for future generations. Others castigated germ-line engineering as the height of scientific, religious, and ethical hubris. The argument against "playing God" has been a recurrent theme in such discussions. Since we cannot (except in the case of a few mutations known to cause severe, uniformly fatal disease) characterize genes as good or bad, how, some ask, can we possibly decide which ones to replace or alter? The "value" of a gene to an organism changes over time, and it may vary significantly depending on the environment into which the individual who carries it is born.

A classic example of the difficulty in valuing most genetic variation is the gene that makes part of the oxygen-carrying molecule called hemoglobin. A person in whom one copy of the gene for β-hemoglobin has a single change in the DNA at a particular site (changing the code for the sixth amino acid in the molecule's β chain from glutamic acid to valine) and in whom the other copy is normal has a condition called sickle cell trait. This is not an illness, and, depending on the environment, can be a tremendous advantage. Persons with sickle cell trait are highly resistant to developing malaria, still one of the world's great killers, especially of children. On the other hand, persons who inherit two copies of the sickle cell allele (one from each parent) are born with sickle cell disease, a serious blood disorder that is debilitating and with which—until recently—affected persons often died young. In North America in the late 20th century, having a copy of the sickle cell gene is of little value; in equatorial Africa it can be life-saving.

Those who oppose germ-line engineering argue that we should not even contemplate the prospect of taking control of our genetic destiny. They liken those who would undertake this goal to the sorcerer's apprentice in Disney's *Fantasia*. To do his work, the foolish little apprentice opened a forbidden book and called forth tools he could not manage. In the mid-1970s, more than a dozen major religious groups issued statements condemning germ-line engineering as unethical, unnatural, and against God's will. Reminders of the banishment from Eden resonated through these documents. Most ethicists also viewed such experiments as inherently wrong, largely because they thought them to constitute unethical experiments on human embryos for whom no valid consent could be obtained. This view was incorporated into government policy in the late 1970s when NIH issued regulations that forbade the use of federal funds to conduct fetal research (see Chapter 23).

The most troubling aspect of germ-line genetic engineering is that in addition to the welcome prospect of "treating" embryos to avoid severe disease in them and perhaps their offspring, it raises the much more troubling possibility of "enhancing" the genomes of normal embryos by adding genes that would increase the likelihood that the children would (to mention just a few obvious choices that might intrigue some parents) be more intelligent, have greater athletic prowess, be musically talented, or have greater beauty than if they were conceived naturally. To some this seems like a violation of the ethics of parenting, for it implies that the couple who chose to use germ-line enhancement must have thought that any child they might have conceived naturally was destined to be inadequate in their eyes. Others argue that nothing is more natural than efforts by parents to give their children the best possible genetic advantage in this uncertain and harsh world.

From the start, the scientific, bioethics, and religious communities and regulatory bodies drew a sharp line to separate discussions of somatic cell gene therapy from germ-line gene therapy. In June of 1983, Senator Mark O. Hatfield placed in the Congressional Record a resolution signed by 56 religious leaders urging that "efforts to engineer specific genetic traits into the germ line of the human species should not be attempted." Even prior to that statement, the federal Recombinant DNA Advisory Committee had taken the position (from which it has never retreated) that it would not approve any proposals to use federal funds for experiments in germ-line genetic engineering.

Although it is impossible to know how deeply the story affected federal policymakers, religious leaders, or the general public, genetic engineering did begin on a most inauspicious note. In the summer of 1980, a decade before the Human Genome Project was initiated, a time when gene therapy was still only a dream at NIH, word surfaced that Dr. Martin Cline, a prominent hematologist at UCLA, had attempted to treat children suffering from β-thalassemia (a genetic blood disease) in Israel and Italy with artificial genes. When Donald Frederickson, the Director of the NIH, heard the rumor, he immediately initiated an inquiry, for if the story was true, Dr. Cline had violated federal rules on research involving human subjects, which in addition to being unethical, could provoke a firestorm of criticism in Congress. It was true.

Earlier that year, Cline had submitted a gene therapy proposal to an Institutional Review Board (the name given to the local committees that the federal rules require must review proposals to obtain federal grants to conduct experiments on human subjects) at UCLA, and the IRB had rejected it. Cline had then arranged to conduct his experiment outside the country. In the fall of 1980, investigators concluded that Cline had misled his colleagues in Israel and Italy about key scientific details of his work, and, in treating the children, had violated federal rules. He eventually admitted that he treated human beings with recombinant DNA molecules without first obtaining the required NIH approval but defended his actions on medical grounds, arguing that the course he took might have saved children from a fatal disorder.

Cline was censured; NIH terminated two of his grants and UCLA forced him to resign as chairman of his department. An editorial likened him to Dr. Frankenstein, arguing that his behavior threatened to erode public support of research. Each event was a serious blow; combined they should have been a knockout punch. Cline had the stamina and courage to persist. Despite his self-caused fall from the scientific elite, he has remained highly productive, publishing more than 100 research papers since 1980. With hindsight, it seems that Cline's worst sin was hubris.

A decade passed. Throughout the world in dozens of academic laboratories and a growing number of start-up biotech companies, scientists tried to develop techniques for gene therapy and use them in animal models. At NIH, despite the serious setback caused by the Cline fiasco, Dr. Anderson, then Chief of the Molecular Hematology branch of the National Heart, Lung and Blood Institute, kept the gene therapy dream alive. During the 1980s, Anderson and his colleagues faced huge scientific, ethical, and political obstacles that had to be surmounted before NIH would approve an experiment like the one Cline had conducted improperly. Anderson's tenacity was as impressive as his intellectual prowess. Time and again he went before the Recombinant DNA Advisory Board to propose, explain, and defend his plan to make gene therapy a reality. When his protocol was finally approved by the NIH on September 7, 1990, one journalist observed that the process had taken "three years, three months, one week, and one hour," surely a record of some sort.

THE FIRST TREATMENT

On September 13, 1990, Anderson's dream became reality. He and his colleagues, a seasoned cancer researcher, Dr. Michael Rosenberg, and a pediatrician, Dr. Kenneth Culver (who went on to direct a gene therapy research center at the University of Iowa), met in a hospital room, and infused genetically engineered bone marrow cells into the vein of a four-year-old girl named Ashanthi. Troubled since infancy by persistent infections, Ashanthi DeSilva had been seen by pediatrician after pediatrician until Dr. Ricardo Sorenson, a physician at Rainbow Babies Hospital in Cleveland finally diagnosed her as having adenosine deaminase (ADA) deficiency, an exceedingly rare form of immune deficiency caused by a defect in a gene on chromosome 20. Fortunately, in 1987 scientists had developed a way to administer the bovine version of this missing enzyme in a way that allowed it to circulate long enough in the body to restore some measure of immune function. It helped, but it was not a cure. Even on the medicine, children with ADA deficiency still get more infections than they should and have to be watched with great care.

Although exceedingly rare, ADA deficiency was an excellent choice for the first attempt at gene therapy. It is a single gene disorder primarily affecting the function of only one group of cells, those that comprise a crucial wing of the immunological defenses. It was also important that patients who agreed to undergo gene therapy for ADA deficiency would be able to stay on their current medicines. Furthermore, there are objective methods, such as cell counts, assays of the level of immune response, and studies of the frequency of infections, that can be monitored. These measures can be compared to methods used before gene therapy and thus give an indication of whether they are helping.

What French Anderson and his colleagues hoped to do was to turn

Ashanthi into a chimera. In Greek mythology, the chimera is a fire-breathing monster with a lion's head, a goat's body, and a serpent's tail. Bone marrow therapy creates a different kind of chimera. The successfully treated patient goes through life with two versions of a particular cell. Most, of course, derive from the germ cells with which he was conceived, but the engineered blood-forming cells are different. Gene therapy is much more subtle than bone marrow therapy. If it works, the patient is a chimera for only a single gene, in Ashanthi's case the gene that codes for adenosine deaminase.

The experimental therapy was simple. The doctors at NIH removed a large number of white cells from Ashanthi's blood. They then set up cultures to which they added chemicals to spur cell division. To these they added an attenuated virus known to invade white blood cells into which had been inserted a cloned adenosine deaminase (ADA) gene. The hope was that the virus would carry the normal copy of the gene into the white cells and that in some of them the new DNA would be taken into the cellular genome. As such cells grew and divided, they would be replicating the new ADA gene as well as the old. The newly divided cells would be infused through a regular i.v. into a vein in Ashanthi's arm.

There were many uncertainties. Would the cells in culture take up the vector? Would the cells that did, divide normally in culture? Would they continue to divide normally in Ashanthi's body? If they did, would they make enough normal enzyme to restore her immune function? Could the cells with the engineered DNA endanger her? The researchers could only guess.

In 1999, nearly a decade after her gene therapy was initiated, Ashanthi is doing well, and she has been joined by more than a dozen other children with ADA deficiency who have been treated with engineered genes. However, there are still many unanswered questions about the world's first gene therapy experiments. For most of the children, there is good evidence that a sizable fraction of the white cells placed in their bodies that are supposed to make ADA are doing so. But the patients continue to need follow-up treatments, and none of the clinical geneticists is yet confident enough in their work to stop the standard therapy with the bovine enzyme. Nevertheless, there can be no doubt that in treating Ashanthi, French Anderson and his colleagues opened the door to a new era. Since their pioneering work, interest in gene therapy has exploded.

THE COMMERCIALIZATION OF GENE THERAPY

About 1993 the rapidly growing biotech industry began to attract investors who were willing to bet tens of millions of dollars on gene therapy companies. In 1994 and 1995 it may have been fair to claim, as some pundits did, that there were more companies devoted to gene therapy than there were patients who had ever participated in gene therapy protocols, but that is to be expected. The major focus in research had to be on the development of safe and efficacious methods to deliver genes to the cells of patients. This first requires extensive work with animal models; otherwise, the NIH would (appropriately) reject any proposal to offer an experimental somatic cell gene therapy to humans.

During the early 1990s, both efforts to develop delivery systems and early experiments to treat animals that had single gene disorders analogous to those in humans yielded impressive results. Among the most dramatic was work with genetic disorders of cholesterol. The Watanabe rabbit (named for the Japanese scientist who discovered its illness) has too few receptors on its cells to process circulating low-density lipoprotein (LDL), a molecule that transports cholesterol. These animals have a genetic form of heart disease. In December 1991, a team at the University of Michigan led by James Wilson reported that it had ameliorated the disease in these rabbits by a combination of surgery and gene therapy. Wilson and his colleagues removed part of the animals' livers, grew the hepatocytes (liver cells) in culture, treated them with a retrovirus to which had been hooked a normal copy of the LDL-receptor-making gene, and transfused them into the portal circulation (the blood system to the liver). After six weeks, the liver cells in these animals were making about 4% of the normal level of the protein, but even that small amount had cut their serum cholesterol by an impressive 30%. This was not unexpected. In most cases one needs only about 10% of the normal levels of a protein to have adequate bodily function.

This success encouraged Wilson to propose, and regulatory watchdogs to approve, an attempt to use somatic cell gene therapy to treat a person with the equivalent human disease. The patient was a 29-year-old Canadian woman who had been born with a severe form of hypercholesterolemia caused by almost complete dysfunction of her LDL receptors that made her incapable of clearing the "bad" cholesterol from her blood.

At times her serum cholesterol was over 1000. Even with current treatments such as diet, cholesterol-lowering drugs, heart medications for the angina that develops from coronary artery disease, and bypass surgery, such patients usually die of heart failure by early adulthood. In June 1992, surgeons removed 10% of her liver and sent it to the genetics lab. The geneticists then broke the tissue into cell suspensions, treated her cells in essentially the same way that they had the rabbit cells, and in three days returned them to her through a special catheter that delivered them directly to the liver. The results were gratifying. The woman's cholesterol level fell 30% without any use of cholesterol-lowering drugs.

During the mid to late 1990s, research in human somatic cell gene therapy grew steadily. As of June 1999, the Recombinant DNA Advisory Committee, which must review all gene therapy proposals conducted with federal funds (which is essentially all proposals), had approved 313 protocols, of which 40 focused on efforts to treat just 15 single-gene disorders. Many others attempt to treat a variety of cancers, AIDS, and heart disease. However, the large number of trials is not evidence of rapid progress. Indeed, Dr. Harold Varmus, then the Director of the NIH, was openly critical of the field and faulted it for shoddy basic science. Despite the huge research effort, several scientific hurdles have been difficult to clear. For example, it remains difficult to construct a viral vector that will deliver donor DNA into the genomes of enough of the patients' cells to show a positive clinical effect. Even when an effect is seen, it is usually transient, which, given the cost of gene therapy, suggests that the potential treatment would be impractical. Furthermore, the risks of gene therapy remain exceedingly difficult to quantify. As a result, each research trial must begin slowly with a controlled and careful escalation of the dose of viral particles that contain the therapeutic DNA. Those risks took on a new dimension in the fall of 1999 when an 18-year-old man undergoing gene therapy at the University of Pennsylvania died four days after being injected with the engineered virus.

THE FIRST DEATH

Jesse Gelsinger was born with an unusually mild form of a rare and often fatal genetic liver disease called ornithine transcarbamylase (OTC) deficiency, a defect in one of the genes that code for enzymes which participate in a biochemical pathway that removes ammonia (a by-product of metabolizing proteins) from the body. His disease was mild for an unusual reason. The causative mutation had arisen after conception and only affected some of his cells. He was in fact a "mosaic," having two different cell lines, one normal and one abnormal. Despite occasional hospitalizations for elevated ammonia levels, Jesse, who lived in Arizona, was doing well, but when his pediatrician informed him about a new research program at the University of Pennsylvania to attempt to treat the disorder by delivering a normal OTC gene (which had been attached to a weakened adenovirus) to the liver, he jumped at the opportunity. He felt a deep sense of commitment to help other children who were much more severely affected with other variants of the same disorder.

Jesse, who was the 17th and last person to be treated under the protocol, underwent his therapy in a radiology suite on September 13. Because the gene vector was intended only to treat liver cells, it was delivered via a special catheter inserted into his groin and threaded through his blood vessels to reach his intrahepatic artery, the major source of blood supply to that organ. Jesse became ill within hours of receiving the maximum dose permitted under the protocol. He spiked a high fever, and over the next 48 hours developed acute, fulminate liver failure followed by adult respiratory distress syndrome. He died on the fourth day.

Of course, the research team immediately reported the death and both the NIH and Penn immediately initiated an investigation of Jesse's treatment as well as those of all the other human subjects who had been treated (no one else had become dangerously ill or died). NIH immediately placed a "clinical hold" on a number of related research trials, including two being conducted by the Schering-Plough pharmaceutical company. The death, the first resulting from gene therapy, received wide attention from the press, sometimes with a sensationalist twist that may help to slow research like this over the next couple of years. For example, the *New York Times Magazine* ran a major story on November 28 entitled, "The Biotech Death of Jesse Gelsinger," which suggested that the whole field needed to slow down and be subjected to more detailed oversight.

Any avoidable death is a tragedy. Surely, those committed to advancing gene therapy and those who must review and approve or reject new protocols will long remember Jesse Gelsinger's death. But the prospects are much too promising to stop. In the summer of 1999, an American company called Vical announced that it had used gene therapy to treat prostate cancer and had seen dramatic reduction in tumor size in all 12 men who were treated. In October, Dr. Victor Dzau, chief of medicine at the Brigham and Women's Hospital in Boston, reported that he had successfully used gene therapy to treat clogged leg vessels. His aim is to do the same for heart vessels.

Given that in September 2000 somatic cell gene therapy will celebrate its tenth birthday, it is fair to say that progress has been slow. By the close of 2000, scientists will probably have attempted gene therapy in fewer than 1500 people, and they will not be able to claim an absolute cure for any of them. However, this limited progress must be put in perspective. Once scientists develop highly effective ways to deliver corrective DNA to target cells, an area where there has been steady progress, the use of this therapy will greatly increase.

I expect that between 2005 and 2010, somatic cell gene therapy will emerge as routine treatment for a variety of single-gene disorders, especially all those currently treated with donated bone marrow cells (there are more than a score ranging from the relatively common thalassemias to the rare Wiskott-Aldrich syndrome, a disease of premature aging). One of the great advantages of using gene therapy is that the work is done on the patient's cells so there are no immunological problems such as those that plague persons needing organ transplants.

What is the likely future of germ-line genetic engineering? As we become more therapeutically proficient, there will be instances where it will make sense to go forward in the clinic. For example, consider a couple facing a 1 in 4 risk of having a child with Tay-Sachs disease (a uniformly fatal brain disease of early childhood) who are unalterably opposed to abortion. The couple might opt to conceive by in vitro fertilization, undergo a test called preimplantation diagnosis, and, if the embryo is destined to have the disease, have gene therapy to correct the defect before the future child is transferred into the mother's womb. Such a therapy alters in a minuscule way the future of the human gene pool. It permits a child who might have died in childhood to grow to adulthood and have children of his own, passing on, as all parents do, reshuffled combinations of his genes. Of them, one will be an artificial, but functional, version of the gene for hexosaminidase A, the enzyme which when dysfunctional causes Tay-Sachs disease. It poses no threat to future generations.

Before 2020, germ-line engineering to cure severe genetic disease in human embryos will be an established therapeutic option. It will, however, be used infrequently. Many couples who know they have a 1 in 4 risk of bearing a child with a severe genetic disorder will continue to use prenatal diagnosis and selective abortion (the course followed today by the vast majority of couples who know they are at risk for having a child with Tay-Sachs disease) or preimplantation diagnosis. Preimplantation diagnosis combines the techniques of in vitro fertilization with highly sophisticated DNA-based testing. Technicians tease a single cell away from an 8-cell embryo and test it for the disease. If the embryo is not burdened with the disorder, it is implanted in the woman.

Perhaps even more exciting is the prospect for treating many fetuses known to have genetic disorders in utero. Although this has not yet been attempted in humans, it is attractive because (1) it could avert the signs and symptoms of disease, (2) fetal cells are likely to be more efficient at taking up and incorporating gene vectors, and (3) immunosuppression will not likely be needed. During the last decade, there have been impressive strides in performing surgery in utero to treat major birth defects. As gene therapy requires merely delivering cells to a fetus, the necessary surgery will be comparatively simple.

The great scientific, ethical, and political debate that looms ahead is not over the use of germ-line gene therapy to treat persons suffering from disease, which is as inevitable as it is welcome. The controversy will surround efforts to "improve" prospects for success in life by making humans smarter, more beautiful, more athletic, or more outgoing, or by improving some other potential in their lives. As the technology to accomplish this will eventually be indistinguishable from that used to treat serious disease, this will happen. However, we know so little of the genetic contribution to such qualities, one can confidently predict it is still many decades from realization. It is still anybody's guess, but it is possible that by 2050 germ-line enhancement therapy might be as common as and no more controversial than cosmetic surgery.

PART
6

DILEMMAS

Genetic Technologies and

Individual Choice

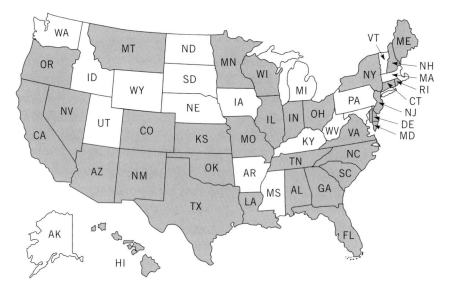

Map showing states having laws to prohibit or control the use of genetic information by health insurers.

21

Genetic Testing and Privacy
Who Should Be Able to Know Your Genes?

Twenty-five hundred years ago, the Greek city-states carried the torch of Western culture. The Greeks worshiped many gods, each with his or her temple and special authority. They believed that one temple, at Delphi, on a beautiful hillside far from Athens, stood at the precise center of the world. Delphi was dedicated to Apollo, the god of knowledge and medicine. For centuries Greek citizens, anxious about life's great questions— Will I marry happily? Will I have children? Will I live to old age?—traveled to Delphi and paid the priests to read the smoke of burning laurel leaves and divine the answers given by Apollo's grace. Early in this century, archaeologists at Delphi unearthed a stone that had lain across the entrance to the main temple. On it were carved the Greek words: "Know Thyself."

Today, genetic testing has supplanted divination, and the Delphic injunction has moved from religion to medicine. We are entering an era where we will be able to find out much more about risks of developing serious illness or of having children with genetic disorders than was even imaginable a decade ago. A decade or two from now our ability to inquire about our genetic risks will be much greater than it is today. What will we want to know? Why will we want to know it? How will the information we learn shape our view of ourselves, our spouse, or our children? Who should be allowed access to genetic information? For what purposes should genetic information be used? Is the threat of genetic discrimination so great that people should forego tests? These questions, and others like them, are the lens that will focus the emerging public debate over the uses of genetic information.

GENETIC TESTING: AN OVERVIEW

Genetic testing began in the early 1960s when we started to screen all newborns for phenylketonuria (PKU), a rare genetic disorder (1:12,000

263

births) that causes mental retardation. By promptly placing affected infants on a special diet which is low in phenylalanine, it is possible to avert otherwise certain retardation. Today, virtually every newborn in the western world is tested for PKU. The program has been so successful that few doctors have ever seen a person who is mentally retarded because of PKU. During the late 1980s I was the medical director of a large residential institution for mentally retarded persons in which lived several adults with untreated PKU. When I arranged a visit by a group of Boston physicians who cared for children with PKU (all of whom had been diagnosed at birth and successfully treated) to observe my patients, the doctors were astounded by the severity of the mental retardation. The persons with untreated PKU, who had no use of language and suffered from seizures, had the tiniest genetic error—just one or two misspelled DNA letters out of billions—but an error with devastating consequences.

Newborn genetic screening, which today has expanded to test millions of children for (depending on the state) three to ten severe but treatable metabolic disorders, has posed relatively little controversy. This is in large part because the tests are highly accurate, and there is much we can do to avert disaster in the children who are born with these rare disorders. However, newborn screening, once limited to a few metabolic diseases, has entered an era that will be marked by dramatic expansion of the kinds and numbers of tests that are run. Currently, the expansion is being driven by the use of powerful, inexpensive technology called tandem mass spectrometry.

Tandem mass spectrometry (TMS) analyzes a tiny volume of blood eluted from a spot on a special kind of paper. The mass spectrometer assesses the size of molecules in the blood, creating a kind of chemical fingerprint. When the machine reports an unusual molecule, the technician (or more likely built-in software) can compare it to a database of metabolites associated with a variety of rare genetic disorders and infer a diagnosis. In 1999 Massachusetts became the first state to include TMS as a routine part of its mandatory newborn screening program. An advisory committee to the Commissioner of Public Health, on which I served, concluded that a disorder known as medium-chain acyl CoA deficiency was sufficiently common and sufficiently treatable to warrant screening. In addition, Massachusetts launched a pilot, voluntary screening program that uses TMS to look for chemical signals indicating that an infant may be affected with one of an additional 20 disorders. About 99% of new mothers who are invited to participate in the pilot program do so.

TMS has arrived at an excellent moment, both because of the benefits it offers and because it is a surrogate for a not-too-distant time when all newborns will routinely undergo DNA screening that may ascertain and compile thousands of questions about their genetic health. A limited DNA testing program is already part of newborn screening in Massachusetts; it is used to confirm the diagnosis of cystic fibrosis in infants with a positive test on a simple biochemical screen. In debating the proper uses of TMS, we will rehearse the future use of DNA testing of newborns. Today, DNA testing is too expensive to use on all newborns, but costs are decreasing rapidly. With the advent of a low-cost DNA array chip (a device on which are anchored thousands of short segments of DNA each designed to look for a particular mutation), we will have to decide what kinds of information it is appropriate for the state to find out about people.

No one has ever challenged the constitutionality of laws that require babies to be tested for these rare disorders. However, if newborn genetic testing expands to identify infants with less severe problems (which it is almost certain to do), it is inevitable that some parents will argue that compelling them to submit their children to genetic testing is an invasion of privacy. And so it is, but the real question is whether a mandatory test confers a benefit to society that outweighs the potential harm to individuals. Historically, the courts have accorded the state significant latitude in the exercise of the "police power" on behalf of the public health. One could argue, however, that programs to identify children at risk for genetic disorders do not seek to find persons who pose a risk to the public health, and there is thus an insufficient basis to compel testing. In addition, one could argue that any test that merely produces information about a degree of risk for an adult-onset disorder or for a non-life-threatening condition falls far short of justifying mandatory testing.

Prenatal testing raises even tougher ethical issues than does genetic screening of newborns. In our society, pregnant women are generally offered two tests aimed at identifying fetuses with birth defects. Those women who will be 35 or over at delivery are advised about the use of amniocentesis to find out whether the fetus has a chromosomal disorder, mainly Down syndrome. Virtually all pregnant women are offered a blood test to determine whether the fetus might have spina bifida (improper closure of the spinal column). More than a third of pregnant women over 35 undergo amniocentesis (aspiration of amniotic fluid to obtain fetal cells), and about 80% of those who learn they are carrying a fetus with Down syndrome decide to end the pregnancy. About three-quarters of pregnant women take the blood test (which measures proteins made by the fetus that cross the placenta into the mother's blood) to screen for spina bifida. About three-quarters of those women who learn that the fetus has open spina bifida (thus making it likely that they will have moderately severe birth defects) terminate the pregnancy. These are painful decisions, not the least because doctors usually cannot predict the severity of the impairment that the fetus is likely to have.

The third major type of genetic screening in use today seeks to identify persons who carry genes for recessive disorders. Should these otherwise healthy persons marry another carrier of a mutation for the same disorder, the couple face a 1 in 4 risk in each pregnancy of having a child with the disease. In the United States, population-based carrier screening has focused on identifying persons at risk for bearing children with sickle cell anemia (mostly African-Americans and people of Mediterranean origin) and persons at risk for bearing children with Tay-Sachs disease (mostly Ashkenazi Jews). Both the carrier screening programs and the outcomes have been dramatically different. Despite more than 20 years of screening, there has not been a reduction in the expected number of births of children with sickle cell anemia. During the same period, the births of children with Tay-Sachs disease in the Jewish community has fallen by about 95%. There are many reasons for this disparity, including differences in access to health care, and cultural differences about the use of selective abortion, but the major one is the difference in the severity of the two diseases. Tay-Sachs disease is a uniformly fatal disorder of early childhood. Persons with sickle cell

anemia have many medical problems, but a child born with the disorder today has a good chance to live a productive life into his or her 40s
or beyond.

Prenatal testing and carrier screening have made it possible for
women at risk to avoid the birth of children with two of the most
common birth defects and two of the more common recessive disorders. For example, in England, prenatal screening and selective abortion have led to a 95% reduction in the birth of children with spina
bifida. Many women now regard genetic testing as an important component of their health care. From a public health perspective, the
picture is more complex. One could argue that western society has
embarked on a course that seeks to avoid the births of persons
with certain kinds of disabilities. Such a decision may seem fairly
straightforward in cases of severe, incurable disorders (e.g., Tay-Sachs
disease), but consider a future in which we will examine hundreds
of genes to find out about a huge variety of risks. Will some women
abort a fetus because it is at high risk in mid-life for ovarian cancer?
Will others want to abort if they learn that the future child will almost
certainly be severely obese? Should tests that merely predict health issues in adulthood even be offered as part of carrier screening or prenatal diagnosis?

As DNA-based predictive tests proliferate, some women and couples
will use them in ways that others will find offensive. In the United States,
each year about 1,000,000 women terminate pregnancies, in more than
98% of the cases because they do not wish to bear a presumably healthy
child. Public opinion, which remains bitterly and intractably divided over
abortion, will surely be inflamed over the growing practice of selective
abortion—the termination of a pregnancy based on a particular prenatal
diagnosis. But does not the right to reproductive privacy include a
woman's decision to terminate a pregnancy even if it is for reasons that
many persons abhor?

For 30 years genetic testing, whether it be newborn screening,
prenatal diagnosis, or carrier screening, has focused on identifying
risks for serious disorders that would be present at birth. The revolution in molecular genetics will change the nature of the data that could
be made available about us. For example, many of the DNA-based tests

that could be used in the future screening programs will not diagnose a disorder. Rather they will recalculate the odds of developing a disorder. In so doing, they will raise new and difficult questions about what we might want to know and what use we can make of such knowledge. The following vignettes illustrate some of the dilemmas we will face.

Taking the Test for an Untreatable Disorder

During the last 10 years, Peter Loomie, a soft-spoken computer programmer, watched his father and one of his uncles die in their 50s of a rare form of Alzheimer disease. In his family, the disease is due to a mutation in a gene on chromosome 14. Peter, who is in his mid-30s, is married but has no children. A DNA test for the mutation is available, and he has spent many sleepless nights wondering whether to take it. He calls the test the most important coin toss in his life. There is a 50% chance that he inherited the mutation from his father, and a 50% chance that he did not.

The test results might redefine his life. If he takes the test and it shows that he inherited the mutation, Peter does not think he would have biological children because he cannot bear the thought that he might pass the mutation to them (a 1 in 2 risk). Because he would likely become ill a decade hence, he doubts whether he would be willing to adopt. Although he is happily married, Peter secretly fears that if he tested positive, his wife might leave him. He admits that he has no basis for this fear. His wife knew about the family illness before she married him, and since the predictive test became available, she has steadfastly maintained that the decision about whether or not to be tested was his alone.

If Peter takes the test and it reveals that he did inherit the mutation, Peter says he would quickly load up on insurance. He and his wife will stop thinking about their retirement years and redouble their efforts to enjoy the present. If the test shows that he did not inherit the mutation, Peter's risk for early-onset Alzheimer disease vanishes (his risk for the much more common form of Alzheimer disease that strikes in old age is still the same as that faced by the rest of the population). The fear that he and his wife struggled with concerning childbearing—a fear of passing on a "bad" gene to kids—would also evaporate. The way they think about jobs, savings accounts, travel, insurance, and myriad other matters would dramatically change.

Peter is at risk for an exceedingly rare disease, but there is also a predictive test for the common form of Alzheimer disease, the type that affects up to one-third of Americans who live into their 80s. About 2% of us—more than five million people—carry two copies of a gene variant called apoE4 which confers a much higher than average risk of Alzheimer disease (see Chapter 19). Those of us who are born with two copies are not certain to develop Alzheimer disease, but are much more likely than average to do so and at a somewhat younger age (mid-60s). The test for apoE4 is inexpensive and accurate. Should doctors use it as a susceptibility test? Currently, they do not. It is recommended for use only as a confirmatory test in people who are already showing signs of possible Alzheimer disease.

How should doctors respond to patients like Martin Dulbecco, a 45-year-old banker whose father developed Alzheimer disease at 64 (and who had two copies of apoE4), and who insists on being tested? Should he be provided with this information? Martin told me that since he already lives each day believing that he shares his father's fate, the test can only help his situation. If he does not have two copies of apoE4 he will conclude that he will not develop Alzheimer disease in his 60s; if he does have two copies, it will merely confirm what he "knows." How tightly should physicians exert controls on access to genetic information? Do physicians have any right at all to deny access to such tests?

The first highly informative DNA-based predictive test—for Huntington disease (HD), a late-onset, untreatable brain disorder that slowly kills—became available in the mid-1980s. Although he had not taken the test, for years Richard Coe, a 37-year-old tennis pro, whose father died of HD, lived as though he had the HD gene. He lived the motto, "Eat, drink and be merry, for tomorrow we die." As Richard tells it, he spent freely and was always deeply in debt. He had a vasectomy when he was 21, had lots of girlfriends, but never a serious relationship, took up hang gliding, and parachuted regularly. He lived on the edge and loved it. Then, he "made the mistake of falling in love." After two years of internal debate, he took the HD test. To his shock, he learned that he had *not* inherited the mutation. "My life was ruined," he recalls laughingly. "I had to get married, reverse the vasectomy, stop hang gliding, pay off the credit cards, and start being polite to my boss." Richard's story shows that there are vast differences in how individuals react to genetic risk and to learning whether that risk could manifest.

DISCOVERING NONPATERNITY

Genetic testing created a dramatically different problem for Ada Boyd, a pediatrician who was caring for a little girl with cystic fibrosis. As part of her interest in how different mutations within the CF gene shape the severity of the disease, Dr. Boyd tested the DNA of the girl's parents. The test showed that the husband did not have either of the mutations that were present in the little girl's DNA. The inference was inescapable. The husband is not the father of the child. To whom (if anyone) should Dr. Boyd disclose this information? Because the couple sought her help together, she believes that she owes an ethical obligation to each person. If she discloses the finding of nonpaternity to the husband, she may threaten the marriage.

Experienced clinical geneticists confide that the inadvertent discovery of nonpaternity is an uncommon, but not a rare, event in their practices. Although there are few good data, population geneticists estimate that unrevealed nonpaternity is present in about 1–5% of families. How is such a finding handled in the genetics clinic? Surveys indicate that more than 90% of clinical geneticists who discover nonpaternity inform the woman, but not the man. These surveys, however, were conducted before the public's now steady exposure to genetics. Today, if the husband were to ask about the implications of his not having a mutation, it would be nearly impossible to avoid telling the truth. Is there a valid ethical difference between not offering the truth and lying? To me the current practice seems difficult to defend and will almost certainly have to be revised as consumers become more sophisticated about genetic testing.

Discovery of nonpaternity can also confound genetic research. I was consulted by a scientist who had been studying a large family burdened with a rare genetic form of cancer. In mapping the culprit gene, he had discovered that one of several adult children could not possibly have inherited the risk for cancer because she was not the daughter of the parent who had passed on the mutation. Telling her that she could not be at risk would lift a great burden from her shoulders, but it could add a new one. If she asked why, the scientist would have either to lie or to reveal that the woman is not the daughter of the man that she has presumed for 40 years is her father. After much soul-searching, the scientists and doctors decided to hide the discovery of nonpaternity. They even sought and obtained the approval of the editor of the journal in which they published their research to alter some genetic facts about the family so that no one could draw the correct inference.

Warning Others

One of the toughest ethical dilemmas that I have faced in clinical genetics began accidentally. One day in 1986, while examining a mentally retarded man with a high fever, I noticed that he had a prominent forehead, simple cupped ears, and large testicles—features associated with a now commonly recognized disorder known as Fragile X syndrome (named for the fact that, depending on how the cells are prepared for study, the X chromosome may look broken under the microscope). Without thinking through the implications of my decision, I ordered a special chromosome test. Two weeks later, I learned that the test confirmed that this 47-year-old man did have Fragile X syndrome, an X-linked genetic disorder. Knowing that this knowledge had important implications for the man's two nieces, both of whom were recently married, I called his sister (their mother) who was his guardian. I told her the diagnosis, and asked permission to talk to her daughters.

The woman was outraged. Without explanation, she forbade me to contact her daughters and slammed down the phone. Over the next month, I tried repeatedly to meet with her, even sending a registered letter. Finally, a social worker who had the sister's confidence figured out what was going on. The woman, who had for 40 years believed that her brother was retarded due to a fall from a crib, had suddenly and dramatically had the view of her brother's illness reshaped. She now had to confront the fact that her brother was retarded due to a gene that he had inherited from their mother. She may or may not have realized that there was a 50-50 chance that she too had inherited the gene (which tends to affect women much less severely) and, that if she had, there was a 50-50 chance that her daughters had in turn received it from her. If so, in each of her daughter's pregnancies they would have a 1 in 4 risk of bearing a son with Fragile X syndrome.

Why would the woman not talk to me? Why had she told the social worker that she would never tell her daughters about Fragile X syndrome? Her reason was crystal clear. Her daughters, she knew, were pro-choice, and would seek prenatal diagnosis and would abort an affected fetus. She, however, was vehemently pro-life. As she said to the social worker, "I will never be party to telling my daughters something that could lead to the abortion of a grandchild!"

If I had not ordered the blood test to make the diagnosis, I would never have confirmed my suspicions and upset the sister's world. But, now I did know, and, given her adamant position, I was the only person who could alert the two young women to their risks, before they became pregnant. Yet, if I did tell them, I would be disobeying the clear instructions of my patient's guardian and violating his right to privacy. The man's sister did not relent and I did not warn her daughters. I hope that before they became pregnant the young women told their physicians about their uncle, and that the doctors pursued that family history.

I do worry that someday one of my former patient's nieces will tell me she has a son with Fragile X syndrome, and ask me why I never warned her of her risk. Could she sue me? Faced with a strict order not to breach confidentiality, there is little chance that I could be successfully sued for failure to warn her about Fragile X. True, obstetricians who fail to warn pregnant women over 35 of their age-associated risk for bearing a child with Down syndrome are regularly sued for "wrongful birth" (see Chapter 8) when such a child is born, but there are two key distinctions. Neither of the nieces ever sought me out for care and, even if they had, I would clearly be under an obligation not to reveal confidential facts about my patient without his (or his guardian's) permission.

As it spreads through medicine, genetic testing will create many dilemmas about sharing sensitive information within families. At first, most people are surprised that this could be a significant problem, but every family, especially every extended family, has its own peculiar dynamics, its share of skeletons in the closet, and its share of relatives who don't get along. Patients will sometimes tell doctors not to disclose facts that the physicians think should be shared. Perhaps to deal with this issue, we will someday have to amend our notion of confidentiality concerning the transmission of genetic information within families. Our ancient principle of patient-centered confidentiality may evolve into a family-centered principle—one that permits physicians to make limited disclosures to a few relatives if the medical issue is significant. This is a subjective standard, one that in practice will vary from doctor to doctor, but I know of no better approach.

Those who adhere to an absolute principle of confidentiality, a position I disagree with but respect, may find my position more acceptable if it is coupled to another. Before they test patients, doctors must tell them that should the test uncover genetic facts of great potential importance to others, their routine practice is to make sure a warning is made. The patient who is offended by this position may then seek testing through a different physician who does not feel so compelled. The only international survey (conducted by my colleague, Dorothy C. Wertz) of clinical geneticists on this issue indicates that many are beginning to rethink confidentiality. Among 2000 geneticists in 37 countries, Wertz found that the majority would tell an individual's relatives about a genetic diagnosis, over the objection of the patient, if the relatives asked. In the United States, clinical geneticists split down the middle on this issue; few geneticists report they would actually seek out and warn the relatives. About half say that they would disclose if they were directly asked.

Physicians as Gatekeepers

Should physicians have the right to refuse to test at-risk individuals who want to be tested? To suggest that a physician could ethically refuse to perform an available genetic test may seem paternalistic, but medicine is replete with instances of physicians denying patients' requests. To cite the most common example, most physicians will not prescribe antibiotics to people with colds that appear to be caused by viruses, despite the fact that many such patients visit their doctors expecting to get such a prescription.

Consider presymptomatic testing for Huntington disease. For a while in the mid to late 1980s, there were less than 20 academically based laboratories where one could be tested. These labs made it their policy not to test at-risk persons who were under the age of 18 and not to test an identical twin unless the co-twin also wanted to know whether he or she had inherited the mutation. The geneticists reasoned that, because knowing whether one had inherited the allele carried with it absolutely no medical benefit, it was better to let each at-risk child grow up and decide on his or her own whether to be tested, rather than have a parent do it. In the case of identical twins, since testing one person diagnoses two, the early view, to which few doctors adhere today, was to test only with the permission of both.

Unless there is clinical benefit associated with the knowledge, geneticists usually balk if parents request that a child be tested because the parents want to know. One especially challenging case I was called about involved a young woman who had learned that she carried a gene that put her at high risk for breast or ovarian cancer. She wanted to test her two daughters, who were six and eight years old. Why? She wanted to use the results to decide whether to have more children. If either child had inherited the normal allele the mother wanted no further pregnancies. If both children had inherited the cancer gene, she wanted to become pregnant again, hoping to have a child without this added risk. Reasoning that the children would not be tested for their own benefit, doctors refused to comply with her request. They were in part influenced by the suspicion that if the woman learned that both daughters were carriers, she might seek prenatal testing and abort a female fetus who had the carrier gene.

Although the vast majority of abortions occur because women do not wish to be pregnant, in the future, a steadily growing fraction will be as a consequence of a genetic diagnosis. What constitutes a sufficiently severe genetic risk to justify prenatal diagnosis and selective abortion? Should the choice belong absolutely to the pregnant woman? The classic debate that focuses on this topic is sex selection. For many years in the United States, clinical geneticists (except in cases of sex-linked disease) refused to perform amniocentesis to determine fetal sex for women who planned to abort if the fetus was not of the desired gender. Sex selection (based on fetal ultrasound and abortion) is widely practiced in some societies, especially in parts of India. Even in the United States, some physicians, driven by respect for patient autonomy, no longer feel that they have the right to refuse genetic testing to determine fetal sex.

Physicians, geneticists, and genetic counselors are acutely aware that they have been given the keys to a library full of powerful information.

They do not want to sit on Solomon's throne. What should they do? As genetic tests become ever more widespread, physicians will seek public guidance on the proper uses of the test results, as is already happening in regard to regulating the use of genetic information by insurers and employers. In the absence of regulations, physicians will be ever more likely to accede to the requests put to them.

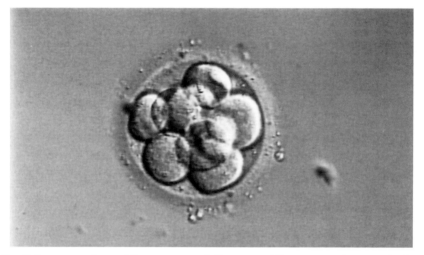

8-cell human embryo. *(Photo courtesy of Sherman J. Silber, M.D., Director, Infertility Center of St. Louis.)*

Frozen Embryos
People or Property?

Davis v. Davis

Mary Sue Davis met Junior Davis when they were both in the army and stationed in Germany in 1979. They married in April of 1980. Mary Sue became pregnant about six months later, but suffered a tubal pregnancy and underwent surgery to remove a fallopian tube. Over the next few years she had four more tubal pregnancies, eventually making her infertile. In 1984 the Davises tried to adopt, but the birth mother changed her mind at the last moment. Mary Sue and Junior then decided to try in vitro fertilization (IVF).

Doctors at the clinic used hormones to control when Mary Sue ovulated and to stimulate her ovaries to produce more than the usual number of eggs. They used a laparoscope to aspirate the eggs, fertilized them with her husband's sperm, and monitored the fertilized eggs (zygotes) as they divided to form a cluster of eight cells. About three days later, they transferred several of these tiny embryos to her uterus. Beginning in 1985, Mary Sue endured six cycles of IVF, at a cost of $36,000 (it would be considerably more today), but she did not become pregnant. Her experience was, at the time, not unusual. Although IVF had produced its first live-born child, a girl named Louise Brown in England, in 1978, progress was slow. Overall, during the mid-1980s success rates in the rapidly proliferating number of IVF clinics were about 15–20% per cycle.

The couple decided to delay a seventh attempt at IVF for a few months until their clinic developed its cryopreservation services. By 1987 many IVF clinics, making use of new techniques that permitted human embryos to be safely frozen and thawed, were routinely creating more embryos than could be used. Research had shown that it was possible to freeze these microscopic clusters of cells for several years, then thaw and implant them with about the same success as freshly created embryos. When aspirating eggs, doctors routinely collected far more than were needed for a single attempt at pregnancy. Cryopreservation provided the option of fertilizing every egg that was aspirated (sometimes as many as 25). One could then implant three or four embryos and freeze the others in case the woman did not become pregnant and wanted to try again.

During Mary Sue's next attempt at IVF, the doctors extracted nine eggs. On December 10, 1988, they transferred two embryos into her womb, and froze seven. She did not become pregnant. In February, 1989, Junior Davis filed for divorce. The only issue that the couple could not work out during the divorce was what to do with their seven frozen embryos. Mary Sue, who had invested so much emotional energy and endured so much discomfort in her fight to conceive and bear a child, steadfastly insisted that she wanted to use them to try to become a mother. Junior vehemently opposed using the frozen embryos in a procedure that would, if successful, confer fatherhood on him as a postscript to a marriage that had ended. They took the matter to court.

At the trial to determine the disposition of the embryos, Junior testified that he did not believe that he would be able to have a healthy parental relationship with a child conceived outside of marriage with a woman from whom he had parted. Junior readily agreed that his views were colored by his own childhood. He was the fifth youngest of six children of a marriage that ended when he was five. Soon thereafter, his mother had a nervous breakdown, and he and three brothers had spent their childhood in an orphanage. He saw his mother only once a month, and he only saw his father three times before he died in 1976.

Junior said that he was terrified of bringing a child into the world who might be destined for a troubled childhood. To his mind, starting out with a set of parents who had divorced before one had started to grow in one's mother's womb was an inauspicious beginning. Junior wanted to destroy the frozen embryos. He was not even willing to donate them anonymously to a woman who could not produce eggs. He felt that no one could guarantee that his genetic child would be born into a marriage that would be nurturing and long-lived.

By the time of the Davis lawsuit there were thousands of embryos in storage, but no court in the world had yet been asked to decide who should get custody of one after a divorce. Only three relevant cases had ever been litigated, two of which involved tangential issues. The first (in 1978) arose when a woman in New York sued Columbia Presbyterian Medical Center after an obstetrician on its faculty intentionally destroyed a petri dish in which IVF was being attempted with her eggs. She recovered $50,000 for emotional distress, and the physician lost his job. The second arose as the Davis case was being tried. A couple sued the Jones Institute for Reproductive Medicine in Norfolk, Virginia, after it refused to send their frozen embryos to a rival IVF clinic in San Diego, California, where the couple had moved. The federal district court treated the matter as a property dispute, and ruled that the Norfolk clinic must release the frozen embryos to their owners.

The first case to confront the uncertain legal status of human embryos arose in the early 1980s when a wealthy Los Angeles couple, Mario and Elsa Rios, died in a plane crash in Chile. In 1981 they had traveled to the Queen Victoria Medical Center in Melbourne, Australia to undergo IVF. Doctors had implanted one embryo and frozen two. When the couple died a year later, the clinic realized that the Rioses had never indicated their wishes concerning disposition of unused embryos, that it had no policy on the subject, and that the law of Australia could provide little guidance. Mr. and Mrs. Rios had no heirs and had left no will. Thus, their frozen legacy raised a number of perplexing issues. What is the legal status of a frozen embryo? If a now-dead couple made no provision for an embryo, does the state have a right to donate it to another couple or destroy it? Should next of kin control the fate of the embryo? Should the clinic? If the clinic donated the two embryos to an infertile couple, would a resulting child have a claim on the Rios estate? Almost certainly not. English common law has long presumed that a child is the offspring of the mother who gives birth to him or her. The Rios embryos were never transferred to an infertile woman, but the case provoked global discussion among bioethicists and others about the need to resolve such fundamental issues.

Questions raised by the Rios matter took on new urgency in March of 1984 when Australia witnessed the birth of a little girl named Zoe Leyland, the first child ever born from a frozen embryo. Several governmental bodies quickly formed advisory committees to develop guidelines concerning the practice of IVF, a task complicated by a clash of views of two major religious groups. The Anglican Church favored the destruction of unused frozen embryos, while the Catholic Church urged that "under no circumstances should they be discarded or destroyed." The Australia Right to Life Association also weighed in, urging that courts appoint guardians for frozen embryos. Australia became the first country to require clinics to ascertain the wishes of couples in anticipation of death or divorce. Unfortunately, the United States did not follow that lead.

The *Davis* case was tried in Knoxville, Tennessee, before a judge who had strong pro-life views. Holding that life was preferable to nonlife, he awarded "custody" of the "children in vitro" to Mary Sue Davis, who had testified that she would try to become pregnant with them. The judge issued an opinion that was sharply defiant of *Roe v. Wade*, the 1973 abortion rights decision which had made clear that a fetus was not a person under the Constitution. It had no chance of being upheld.

I became involved in *Davis v. Davis* on appeal. Acting as friends of the court, a number of groups, including several national organizations of physicians and scientists, were preparing appellate briefs to challenge the scientific foundation upon which the judge had built his reasoning. As "friends of the court" (parties not directly involved in the case, but which had an interest in the issues), their main goal was to clarify clinical and scientific facts about early human embryology. I was asked by the Board of Directors of the American Society of Human Genetics (ASHG) to review a brief prepared on behalf of the American Medical Association that explained the biological nature of an eight-cell embryo, and asserted that there was no scientific basis to characterize it as a child. To me this seemed straightforward on scientific grounds, however debatable it might be on moral grounds. An eight-cell human embryo has no organs and no nervous system; it feels neither pain nor any other sensation. At this stage of human development the fertilized egg has divided three times. Under the microscope the cells look like a tiny mass of soap bubbles. Each cell is probably totipotential; that is, if separated from the others, it can restart the process of becoming a human, which is how identical twins come into existence.

I approved the brief, and the ASHG added its name to the list of signers. The document was one of several that (hopefully) helped the appeals court to articulate its disagreement with the trial court. Sharply rebuking the trial court judge, it characterized the frozen embryos not as children, but as a special form of property. Deciding that Junior Davis should not be forced to become a parent against his will, it awarded full control over the embryos to him, realizing that he would destroy them. Mary Sue immediately appealed.

It took two years for the case to reach the Supreme Court of Tennessee, and by that time Mary Sue's life had changed substantially. She had remarried, and no longer wished to become pregnant with the frozen embryos created with her former husband. Yet, she still sought custody of them, but now it was to donate them to an infertile couple who could not use their own eggs. Junior still wished to destroy the embryos.

The Supreme Court of Tennessee was unhappy with both the trial court and the appellate court opinions. As the U.S. Supreme Court had (in deciding *Roe v. Wade*) clearly refused to confer personhood on a fetus, the Tennessee high court held that the trial court had no constitutional basis for conferring personhood on human embryos. It also rejected the decision by the lower appellate court that the frozen embryos were property, however special. It held instead that unimplanted embryos were in an "interim category" that entitled them to special respect because of their "potential human life," but that they did not have full human life. The court's opinion is, of course, in sharp conflict with the view of many Americans, especially Catholics, whose faith holds that human life begins at conception.

The Tennessee high court had to decide which parent would be awarded control of the embryos and provide a rationale for its decision. Its work was made considerably easier by Mary Sue's acknowledgment that she no longer sought to have the embryos implanted in her. In its June 1, 1992 decision, it took the view that, "Ordinarily, the party wishing to avoid procreation should prevail, assuming that the other party has reasonable possibility of achieving parenthood by means other than use of the pre-embryos in question." It awarded control of the seven Davis embryos to Junior, who promptly directed that they be destroyed. The only woman justice on the Tennessee high court, Martha Craig Doherty, wrote the opinion.

A single state supreme court decision is the beginning, not the end of the effort to craft reasonable legal rules. It was obvious at the time that many similar cases were bound to arise, and so they have. The second was decided in January, 1995, when a New York trial court, faced with facts quite similar to the *Davis* case, reached a dramatically different conclusion. In 1993, after years of unsuccessful attempts to achieve pregnancy, and just two months after she and her husband had created, frozen, and stored five embryos, Maureen Kent Kass filed for divorce. She wanted, nevertheless, to continue to attempt pregnancy via IVF. Her husband sought to have the embryos donated for medical research. The trial judge, asserting that he was required to do so by the rules laid down in *Roe v. Wade*, held that Maureen Kass should have control over them. Since *Roe* held that a woman has the right to end her pregnancy despite her husband's wishes, the judge reasoned that the same logic should apply here. In his words, "From a prepositional standpoint, it matters little whether the ovum/sperm union takes place in the private darkness of the fallopian tube or the public glare of a petri dish. To deny a husband rights while an embryo develops in the womb and grant a right to destroy it in a hospital freezer is to favor procedure over substance."

Maureen Kass promptly announced that she would seek to undergo IVF as soon as possible. Further fanning an emotional conflagration, her lawyer, Vincent F. Stinple, stated that if his client became pregnant through IVF and bore children, she would sue her former husband for child support. The case was appealed to the highest court in New York, which rejected the trial court decision and refused to permit Maureen to have the embryos implanted into her or donated for research.

Similar disputes are now winding their way through the courts of Illinois, Massachusetts, Michigan, Texas, Alabama, and New Jersey. The Massachusetts case, which was decided by the state supreme court in February, 2000, also arose out of a divorce. Despite the divorce, A.Z., the ex-wife, wanted to use two frozen embryos to attempt to become pregnant. At different times she had obtained her husband's written consent to give her control of the stored embryos should they end the marriage. Nevertheless, in the probate court the judge found in favor of the husband who wished to block A.Z. from using the embryos. The judge was impressed that A.Z. was already the mother of twins born after a successful IVF procedure and that the divorce represented a dramatic change in circumstances between the parties to the original agreement that nullified the earlier consents. When A.Z. appealed, the Supreme Judicial Court transferred the case to its own docket. It affirmed the lower court's decision to grant a permanent injunction in favor of the husband that forbids A.Z. from "utilizing" the embryos. According to John Robertson, a University of Texas law professor who is the nation's leading scholar on this issue, the legal rule that is emerging is that "the party wishing to discard wins, unless there is no other way for the party seeking implantation to reproduce."

Given the high frequency of infertility, the use of cryopreservation, and the high divorce rate, we can expect to see more legal battles over embryos, especially since few legislatures seem willing to craft the needed regulations. This reluctance is almost certainly due to their terror of enacting any law that seems to support the holding in *Roe v. Wade*. Although more than 100 IVF clinics are operating in the United States, more than 20,000 IVF babies have been born, and at least thousands of human embryos are in "cold storage," the industry remains unregulated. In sharp contrast, in Great Britain since 1990 the Human Fertilisation and Embryology Authority has licensed and monitored all IVF clinics, keeping track of stored embryos and enforcing rules on disposition.

Surveys indicate that only about 10% of Americans who have stored embryos will eventually donate any of them to other couples. Based on the growing number of couples choosing IVF, the fact that most couples undergoing IVF do store embryos, and the fact that IVF pregnancy rates are rising, there may already be more than 100,000 human embryos frozen in vials of liquid nitrogen in the United States. The number of stored embryos will continue to grow here and throughout the world, for only a few nations require their destruction. Among those that do regulate storage, Australia limits it to ten years, Great Britain, Canada, and France restrict it to five years, Norway to three, and Sweden and Denmark to just one year. Germany permits destruction of the embryo only with the consent of the couple. On the other hand, Brazil permits indefinite storage and forbids destruction.

In 1999 Louisiana was the only state in the United States that had a statute on the disposition of unwanted frozen embryos. It forbids their destruction and requires that those that are unwanted be placed for "adoptive implantation." Although it has not yet been constitutionally challenged, the statute, which seems to accord the status of children to the embryos, would almost certainly be overturned as a violation of the privacy rights of the genetic parents. In the rest of the United States, storage policies are set by individual IVF clinics. Few, if any, will donate an unclaimed frozen embryo without having written permission from the couple who provided the gametes. Today, most clinics require couples to specify in advance how they will dispose of unused embryos.

In the summer of 1996, the British Human Fertilisation and Embryology Authority provoked an international debate when it announced that, pursuant to the 1990 law, it would destroy all unclaimed frozen human embryos that had been in storage for more than five years unless the genetic parents requested that they be stored yet longer. Many of the 910 couples known to have embryos in storage could not even be located. Of those who were contacted few responded, presumably because they had completed their families or had given up trying.

For a week the issue dominated the British press. Letters to the editors compared the plan of "mass destruction" to the Nazi death camps. The Vatican newspaper, *L'Osservatore Romano,* called for thousands of Italian couples to volunteer to have the embryos implanted. A few women did volunteer. A group of Italian physicians offered to pay for shipping the embryos to Italy for implantation into infertile couples. This campaign ended when the British Human Fertilisation and Embryology Authority ruled that it was unethical to donate the embryos without the consent of the genetic parents. On August 1, 1996, the staff at 30 infertility clinics destroyed 3300 frozen embryos, in most cases by thawing them and pouring them down the drain. This drama of mass destruction will not recur in Great Britain. As of August 1, 1991, all couples who store embryos must first agree that unimplanted embryos will be destroyed after five years unless they specify another course of action.

The 3300 "unclaimed" embryos constituted only about one-third of the total that had been stored in Britain. About half of the genetic parents of the other 6000 have donated their embryos to research (which is legal in Britain during the first 14 days of development), 30% are retaining them for possible future efforts to bear children, 10% have ordered them to be discarded, and 8% have donated them to other infertile couples. Surveys suggest that the donation rate is low because couples fear that it will bother them to know that a genetic child of theirs exists but has no contact with them. Some also fear that their own children would be upset to learn of the existence of an unknown genetic sibling.

RESEARCH

Societal debate over the creation and/or use of human embryos for *research* is even more contentious than the disagreements over storage and donation. During the 1980s, blue ribbon committees from about a dozen countries issued white papers on embryo research. Britain's 1984 Warnock Report (named after the Chairperson, Mary Warnock) was most influential. Convinced by the arguments that the study of embryos was essential to understanding infertility, normal development, and birth defects, the Warnock Committee proposed a rule that was favorable to researchers. It recommended that "research may be carried out on any embryo resulting from in vitro fertilisation, whatever its provenance up to the end of the fourteenth day of fertilisation ..." but, that it should be a "criminal offense to handle or to use as a research subject any live human embryo derived from in vitro fertilisation beyond that limit." It also strongly recommended that trans-species fertilization involving human gametes, placing a human embryo in the uterus of another species for gestation, and the sale or purchase of human embryos should be criminal offenses.

When it picked 14 days as the upper time limit on performing research on unimplanted human embryos, the Warnock Committee drew a line firmly in the sand. Its reasoning was clear; 14 days is the time at which one can recognize the existence of a primitive body plan—when the embryo begins to look human. At that time the embryo, which is not much bigger than the period ending this sentence, begins to take on attributes of humanhood. In their view, this was the right point at which to forbid its further use as an object of research.

During the Reagan era, no federal funds were permitted for use in studying the process of in vitro fertilization, let alone embryo research. When Clinton entered office, he quickly lifted the most onerous ban on embryological research. Soon thereafter, a 19-member panel was convened to advise the National Institutes of Health on embryo research. After a year of deliberations, the panel, chaired by Steven Muller, former president of Johns Hopkins University, concluded that the NIH should fund research on the human embryo for the first 14 days of development so long as the studies were carried out under stringent guidelines that protected the embryo from being subjected to what many would consider highly offensive experiments. Presciently (see Chapter 23), it ruled out research that involved deliberate twinning and cloning. It also voted against the creation of hybrids between humans and other animals, the attempted transfer of human embryos into the wombs of animals, and efforts to develop artificial wombs.

The Muller report urged the end of a 15-year era in which no human embryo research had been funded by the federal government, but it quickly ignited a controversy. The National Conference of Catholic Bishops and the American Life League both promptly opposed it. The most offensive issue was the panel's advice that scientists should be permitted to use donated eggs and sperm to create human embryos in the lab, study them, and destroy them at will. President Clinton was quick to order Harold Varmus, the director of the NIH, not to release funds for such research.

Clinton knew that many in Congress were angry because in 1993 a scientist at George Washington University (which sits virtually in the congressional backyard) had performed research with human embryos that violated NIH guidelines. Jerry Hall, a researcher in the lab of Dr. Robert Stillman, had disclosed that, working with early embryos that had been rendered incapable of normal development (embryos that he had obtained as discards from an IVF clinic), he had succeeded in creating an army of potential twins. Hall managed to tease apart 17 human embryos at either the two-, four-, or eight-cell stage and coaxed the isolated cells to grow into 48 clones. When the *New York Times* investigated the research, it came to light that Hall had never obtained approval from a human studies committee (usually called an institutional review board or IRB) to do the experiment. Internal review at George Washington University eventually concluded that the researchers had violated NIH rules. Late in 1994, university officials ordered the two to destroy all the records of their work and to publish no further papers based on the research (a measure that recalls the debate that began just after World War II about whether mankind should try to benefit from knowledge gained through unethical experiments conducted by the Nazis).

The twinning experiment, which applied a technology widely used in animal husbandry to human embryos, generated immense interest in the media. *Time, Newsweek,* and *US News and World Report* all covered it, as did virtually every major newspaper and news network. A Time/CNN poll revealed that "three-quarters of Americans opposed human cloning, with about two-thirds believing it is against God's will."

The advice of the NIH advisory panel on human embryological research was lost in a blizzard of conflict. In August 1995, the House of Representatives passed an appropriations bill that forbade using federal dollars for fetal research. As a result, little significant research in early human embryology occurs in the U.S. The major centers are in Britain, where it is routine for researchers to be able to acquire fresh material from recent abortuses and to deliberately create embryos for research. In addition to making sure that the woman is undergoing an abortion without a scintilla of coercion and that her consent to donate the tissue is fully informed, the British also require that the physician performing an abortion not be directly involved in the research. The scientists are using the retrieved embryos to study how organs are formed. They are also studying the earliest embryos to find out which genes are active in which cells at which times.

The 14-day rule first set down by the Warnock Committee in 1984 has been adopted in many other countries. Unfortunately, it forecloses many important areas of research that could improve our woefully inadequate understanding of human development. If, for example, a more lenient rule that allowed researchers to study embryos up to the end of the 28th day was adopted, scientists would be able to study the formation of the neural tube, the tissue that becomes the spinal column (a process that is largely complete at about days 21–22 when the embryo is about three millimeters long). The failure of the tube to round out and close properly leads to spina bifida, one of the most common birth defects. There is now a widely used prenatal test for detecting spina bifida in fetuses at about week 15 of pregnancy, and most women who learn that they are carrying an affected fetus obtain an abortion. Research on neural tube closure might yield new insights into the environmental and genetic factors that interact to cause this problem. In turn this might result in new approaches to prevention that would improve upon the current recommendation that all women planning to get pregnant take 400 micrograms of folate (a B vitamin which cuts the risk of spina bifida in half) every day. This would then sharply reduce the number of abortions for this condition.

Some people who reluctantly accept the importance of conducting research on embryos nevertheless strongly oppose creating them intentionally for that purpose. They argue that research should be performed only with unwanted frozen embryos. Unfortunately, this choice would make it impossible to study fertilization itself, defects in which may be key to a significant fraction of the infertility that burdens 1 in 6 couples. Frozen human embryos have a potential for human life that commands respect. No one, including those who provided the gametes that created them, should have the right to subject them to macabre experiments or misuse them in other ways. However, those who regulate research in this area should not consider the 14-day rule to be set in stone. We need a somewhat larger window in which to do this important research.

Dolly, the first cloned mammal. *(Reprinted, with permission, from* Nature *Feb. 27, 1997 © Macmillan Magazines Ltd.)*

Cloning
Why Is Everyone Opposed?

DOLLY

The Roslin Institute, one of the world's leading centers for research in animal husbandry, is housed in a neat collection of buildings near Edinburgh. Surrounded by gently rolling, impossibly green hills dotted with countless puffs of fleecy white sheep, the place seems an unlikely setting for grand, scientific events. The setting seems more that of a highly successful, modern farm than the high-octane scientific laboratory that it is. Yet, it was here in 1996 and 1997 that a scientific team led by Ian Wilmut, a mammalian embryologist, in collaboration with a team at a neighboring biotech company called PPL Therapeutics, cloned a sheep named Dolly, which overnight became the most famous animal in the world.

In creating Dolly, Wilmut and his colleagues repealed a fundamental biological law: that it is not possible to produce a genetic copy of a mammal from a cell taken from an adult. Dolly was not created via the union of sperm with egg. She was created from the genetic material of a single cell—taken from the mammary gland of a six-year-old Dorset ewe. Of course, a sperm and egg did come together to create Dolly's genome. Her genetic parents are the parents of the Dorset ewe from which she was cloned. Dolly was created in a glass dish when the nucleus of the cell taken from the Dorset sheep was inserted into an ovine egg and transferred to the uterus of a surrogate mother belonging to yet another breed.

On nature's mighty evolutionary tree, the human twig arises from the same branch as does the sheep twig. Since Wilmut could clone sheep, there is a good reason to think that we will eventually be able to clone humans. Is there any more dramatic moment in the history of our species than the discovery that we can circumvent the mechanisms of genetic recombination and reproduction by which we evolved? Unless there is some as-yet-undetected aspect of human biology that forbids it (which is unlikely), cloning, in combination with gene modification, will eventually launch humans on a scientific and cultural voyage of immense consequences. At some still far-off time, we will truly be able to guide our own evolution. In some respects, Wilmut's experiment is as dramatic as was the proof that the earth revolves around the sun.

All humans and—until Dolly—all mammals begin life as the union of a single sperm and egg. That first cell, the zygote, contains in its nucleus a full set of all the genes needed to create an adult animal. After the first cell division, the daughter cells of the zygote are still totipotent; that is, they are each capable of producing a human. We know this because of the existence of identical twins. Scientists had long thought that totipotency faded quickly. How else could the body plan be established or organs formed? As the first two cells divide to become four and then in turn a cluster of eight and so on, until after three weeks they have organized as an embryo composed of billions of daughter cells, each cell somehow must be reprogrammed to carry out defined tasks—to be a heart cell, a kidney cell, a brain cell. Scientists had good reason to infer that the price of this exquisite specialization was the loss of totipotency. They thought that the genetic programming that must occur to become, say, a heart cell was much too complicated to ever permit such a cell to retrace its history and reconfigure itself with the set of directions needed to create an entire embryo. Although the nucleus of a heart cell has all the genes that make up a complete human genome, most scientists were convinced that many, if not most, of those genes were permanently inactive.

One of the first challenges to the notion of the fixed nature of a cell's genetic program came in 1952 when scientists removed a nucleus of a cell taken from an early frog embryo and transferred it into a frog egg from which all genetic material had been removed. To everyone's surprise, the genetically engineered egg grew into a frog! In 1975 John Gurdon, a British embryologist, repeated the experiment with a nucleus taken from the skin cell of an *adult* frog. The cloned cells consistently grew into tadpoles, but stubbornly refused to go further. Gurdon's experiments strongly suggested that animals could not be cloned from adult cells. In subsequent years, few scientists decided to risk their careers by assaulting this barrier.

During the late 1980s, several research teams did clone sheep and cows by using nuclear material taken from blastomeres (a cluster of cells from which the embryo arises). This is a different kind of cloning, more akin to the deliberate creation of identical twins. One can think of it as "horizontal" cloning, because the replication event (the creation of the twin) does not involve copying a genome from a parent—an animal which has lived or is living independently. Rather, a cell mass is disaggregated, and multiple copies of the animal that would have arisen from the blastomere are created.

In 1996 Wilmut and his colleagues showed that they could use DNA from a fetal cell, which was sufficiently old that virtually every reproductive biologist thought it had completed the path to irreversible commitment, as the nuclear material for cloning sheep. Their experiment, which gained relatively little attention, was based on a new idea that set the stage for creating Dolly. Wilmut guessed that the secret to successfully cloning mammals from adult cells lay not in the genes but in the cytoplasm which surrounds the nucleus. Drawing on their knowledge of the cell cycle, he and his colleagues induced the cells to enter a quiescent phase (a sort of molecular hibernation during which the DNA would be inactive) which they thought would lead to a loosening of the fixity of the nuclear program. Success with fetal sheep cells emboldened them to repeat and expand their experiments.

What did they actually do? They used hormone injections to make Scottish Blackface ewes superovulate, an intervention quite like the treatment of infertile women to harvest eggs from them for in vitro fertilization. Using a laparoscope, they retrieved the eggs from the ovaries of the Blackface ewes. Next, they placed the eggs under a microscope and used incredibly thin pipettes to puncture them and, quite literally, sucked the nuclei (a tiny sac containing all the genes) out of the cells. They then isolated a nucleus (containing the full complement of DNA) taken from a culture of cells prepared from a biopsy of a physically distinct breed of sheep (the Dorset) and used a mild electrical shock (which creates small holes in the cell wall) to coax that nucleus to move into the genetically empty egg. They cultured the resulting clones for six days, carefully monitoring them for any sign of disorderly development. They then transferred all the apparently healthy cloned embryos into Scottish Blackface ewes that had been hormonally prepared for pregnancy.

They conducted experiments with three populations of cells: some from a 9-day embryo, some from a 26-day fetus, and some from the mammary gland of a six-year-old ewe. Altogether, they created 385 potential clones by fusing enucleated eggs with embryonic cells or fetal fibroblasts and 277 potential clones from cells created by fusing enucleated eggs with adult mammary epithelial cells. Most ewes did not become pregnant, and of those that did, the vast majority miscarried. Altogether, the host mothers gave birth to only eight lambs. Of these, seven derived from cells that had been created with sheep embryos or fetuses, and thus represented a validation of earlier experiments. Of the 277 fused cells that were created from adult Dorset mammary cells, 29 morulas (pre-embryo cell masses) were put in the surrogate mothers, 13 ewes became pregnant, and one gave birth—to Dolly. Being the genetic clone of a Dorset ewe, Dolly looks nothing like her Scottish Blackface birth mother.

Does Dolly have a genetic mother? Does she have a father? Yes. From a genetic perspective, she is the offspring of the pair of sheep that gave birth to the Dorset ewe from which the cell was taken to launch her. That is, she is the genetic child of the two sheep that are the parents of the animal that provided the cell from which she was cloned (animals that from a generational perspective we would call her grandparents). But Dolly is also the identical twin sister of the sheep which provided the mammary cell that became the source of her genetic constitution. Of course she is identical only from a genetic perspective, not from an environmental one. After all, she was born nearly seven years after her older twin!

This raises the fascinating question of Dolly's age. Is she a yearling or is she (in terms of life expectancy, fertility, and risk of disease) an old sheep? We have learned that just as do whole organisms, cells too have life expectancies, usually measured in terms of expected number of cell divisions before senescence. Dolly developed from a relatively old sheep cell. Determining her biological age—best done by pampering her for the next decade—will be of crucial importance. Early studies suggest (but do not prove) that Dolly will have a shorter life expectancy than would a lamb born on the same day.

Reactions to Dolly

As any literate adult who was on the planet in March of 1997 knows, Dolly's birth elicited an unprecedented public response. Suddenly, Ian Wilmut, a mild-mannered fellow, was the best-known biologist in the world. The Scottish lowlands were besieged by the world press, and not a few vans carrying television crews probably shouldered sheep transport lorries off the narrow roads about Roslin. To the furthest corners of the globe, nearly everyone who heard the story seemed to sense that humankind had acquired a vast new power, and many worried that we might not be able to handle it.

News of Dolly triggered two quite different stampedes. One group—politicians, consumer activists, bioethicists, religious leaders, and lawyers—rushed to formulate policies that banned human cloning. The other group—reproductive biologists, scientists who work in animal husbandry, and investment bankers—rushed to conduct other cloning experiments, mostly involving farm animals. President Jacques Chirac of France promptly announced that he would ask the Group of Seven industrialized countries to ban human cloning. He convened a panel of experts to discuss "fears" and "fantasies" raised by the prospect of human cloning. Wolfgang Fruchwald, president of the German Research Association, urged Chancellor Kohl to seek a worldwide ban; the Chinese Academy of Sciences announced it had banned cloning; and in Geneva, Dr. Hiroski Natajima, Director General of the World Health Organization, called human cloning an extreme form of experimentation that should never be done because it violates the dignity of the persons born thereby. President Clinton declared a moratorium on federally funded research relating to human cloning and asked the National Bioethics Advisory Commission to advise him within 90 days. In an unusual twist, Ian Wilmut's appearance before Britain's Parliament (which had a few years earlier enacted a law that has been interpreted to prohibit human cloning) to oppose human cloning was countered by the testimony of Ruth Deech, of the Human Fertilisation and Embryology Authority, who said she could envision rare circumstances in which it might be ethically defensible.

Nicholas Coote, assistant general secretary of the Roman Catholic Bishops Conference, asserted that every human being has the right to two biological parents. He was quick to note, however, that human cloning would not contradict the Church doctrine of unique creation. Otherwise, what moral status could be accorded to naturally arising human identical twins? Jeremy Rifkin, a professional critic of genetics, called for criminal penalties for human cloning and advocated longer prison terms for cloners than those handed down to convicted rapists. At MIT, one of the few places that seemed to find some humor in cloning, an April issue of the student paper announced that the university would "use new state-of-the-art cloning techniques to create a race of super beings." The paper also reported that a "superior" university administrator would be the source for cloning future administrative clones. In a news interview, he proclaimed himself the "prototype" who would "be in charge of all the others."

From Denmark to Brazil, scientists rushed to replicate and extend Wilmut's work. Because of the potential economic benefits that could flow from increasing the average milk and meat yield of herds, many scientists attempted to clone cattle. A little more than a year after Dolly, Jose Antonio Visinton, a veterinary scientist at the University of Sao Paulo, announced he had successfully used embryo splitting to clone cows. Rather than replicate a single cell, he had divided early cow blastomeres and transferred them to surrogate mothers, setting the stage for the birth of a small herd of identical twins. This "laboratory twinning" is technically easier to perform than the nuclear transfer technology used to create Dolly.

On July 24, 1997, just 5 months after reporting the arrival of Dolly, the teams at Roslin and PPL announced the birth of five lambs cloned from fetal cells into which had been inserted a single human gene (they did not disclose which one for proprietary reasons, but it is almost certainly a gene the protein product of which is crucial to treating an important human disease). Studies of one lamb, named Polly, showed that she had incorporated the human gene into her cells. Even if the gene product is not well expressed, the fact that it is possible to insert human DNA randomly into the genome of a fetal sheep cell and clone an animal from that cell which develops normally has immense implications for the development of new therapies. The birth of Polly is an important milestone in refining methods for the production of transgenic animals (see Chapter 14)—the pharmaceutical factories of the future! It also brings the dream of xenotransplantation (see Chapter 16) a big step closer to reality.

In August of 1997, ABS Global, Inc., a small biotech company in Wisconsin, announced that it had developed a highly efficient approach to cloning animals and that ten cloned Holstein cows would soon be born. The announcement came hard on the heels of one that ABS had made just a day earlier of the birth of the world's first calf created via somatic cell nuclear transfer (e.g., as was Dolly), a male named, appropriately, Gene. The ABS approach capitalized on technical improvements developed by a

Florida scientist named Dr. Steen Willadson, who reasoned that if one turned an adult cell into an embryo cell and then cloned the embryo cell, many more pregnancies would result. In this double cloning process, the Wisconsin group first created a clone in the way Wilmut had, then it sheared cells from the resulting blastomere and again fused them with enucleated eggs. After they began to form an embryo, they were transferred into surrogate mothers who delivered them. Whereas the technique used by Ian Wilmut yielded about one live-born clone for every 60 attempts at pregnancy (1–2%), ABS claimed a success rate approaching 40%.

Besides agricultural improvements, what are other potential uses of cloning in mammals? Two important ones are preservation of endangered species and making progress in xenotransplantation (see Chapter 16). In 1999 a rare African cat was born after in vitro fertilization and transfer into the womb of a domestic house cat. The feline surrogate mother doted on her unexpected kitten, despite the lack of any family resemblance. Can cloning be far behind? There are a number of offbeat potential uses of cloning, such as the suggestion from a cat fancier that cloning would permit one to immortalize pets. This is, of course, not the case, but it might well provide a grieving master with a cloned pet that he or she could not distinguish from its deceased genetic twin.

Moral Issues in Human Cloning

The overwhelming reaction around the world to the arrival of Dolly has been that nuclear transfer technology (as cloning is termed in the scientific literature) should *not* be used to create human beings. Sifting through dozens of moral pronouncements, however, one is hard pressed to find rigorous arguments. Rather, there is a widely shared moral revulsion to the notion of cloning humans. What is the source of that revulsion? It seems to be a deeply held belief that every human should have the right to begin life (be conceived) with a unique genetic constitution. For example, the first clause of the statement on cloning by the European Parliament says that "each individual has a right to his or her own genetic identity and that human cloning is, and must continue to be, prohibited." What then of identical twinning (which occurs about once in every 270 births)? Are twins born bereft of a fundamental right? Identical twins are *natural* clones, the product of a normal conception about which the couple (if they thought about it) assumed that the resulting child would be genetically unique. Thus, identical twins, whether hoped for or not, are never *intended* because (for now at least) humans attempting to have children cannot act in a manner that improves the chance of achieving that aim. On the other hand, intentional human cloning (whether by splitting an embryo to create octuplets or by nuclear transfer technology) would knowingly and deliberately violate that natural right.

What precisely is the harm to one who is created by a method that denies genetic uniqueness? Identical twins are genetic replicas of each other. Are they in some fashion diminished? How? Psychologists have been studying identical twins for a century and have found very little evidence that the experience is harmful. Quite to the contrary, identical twins typically have a close, supportive relationship, a wonderful asset in our troubled world. Simply put, there is no evidence that one is harmed by being a twin.

Indeed, the argument that one is harmed if one is created in a manner that precludes genetic uniqueness is based on a false assumption: that we are unique because of our genes. In addition to being scientifically erroneous, such an assumption is a capitulation to genetic determinism and a dangerous political idea. Scientist or not, our life experiences should convince each of us that we are the product of an infinitely varied set of interactions with our world. All organisms are at any given moment the embodiment of the history of the interactions of their genetically programmed selves with the environment.

Furthermore, from a genetic perspective, one could even argue that a cloned individual would have a richer and more varied genetic constitution than any individual conceived naturally. Each of his or her cells actually would have genes from *three* individuals. The nuclear genome would be, like everyone else's, the result of a usual conception between germ cells from two individuals (in this case the parents of its "older twin"). In addition, he or she would inherit mitochondrial DNA (small circles of DNA coding for a few key genes) from a third individual—the donor of the egg into which the nucleus was transferred! Thousands of copies of the circular mitochondrial DNA would be in the cytoplasm of that enucleated egg. Finally, because the genome from which the clone was created existed for years in a somatic cell, it almost certainly acquired a number of mutations that differentiate it from the zygote from which it derived.

The strongest objections to human cloning seem to arise from those with the most fertile imaginations. The potential for some of the imagined abuses is real; for others, however, there is no basis for concern. For example, fears that some person or groups will create battalions of genetic copies of individuals for personal gain or for a destructive goal have no grounding in reality. Cloning requires access to human eggs and women willing to act as surrogates, both of which are and will continue to be scarce. It is possible that scientists will eventually discover a way to modify enucleated cow or pig eggs in order to use them as a suitable home for a human somatic cell nucleus. The scientists would also have to figure out either how to transfer human mitochondrial DNA or convert animal mitochondria to meet the needs of human cells, but this too may be a surmountable challenge. Assuming technical success, this would eliminate one major hurdle to inexpensive cloning. However, it is difficult for me to imagine any regulatory body in any nation permitting the experiments that would have to be undertaken to pave the way for clinical attempts to use animal eggs. With enough money, one can, admittedly, acquire even such precious services as those provided by surrogate mothers, but one could hardly do so in secrecy. Surely, global moral censorship would dissuade all governments and all corporate entities from attempting human cloning, but there are no certain mechanisms to prevent isolated acts by a few individuals.

Discussion of technical issues may obfuscate the key point. If the most fundamental fear is that we are entering a brave new world in which we will be able to replicate human genomes, then we must make sure we grasp the reality of being a clone. Even if one made many copies of an individual and arranged to raise a dozen clones, it is highly doubtful that any of them would wind up living a life substantially like his or her genomic parent. True, clones of Michael Jordan might tend to be tall, handsome, and athletic, but as each would grow up in a profoundly different environment from the one that shaped him, it is impossible to predict a life scenario. Could one of the clones become a basketball superstar? Possibly. But the 25-year wait to find out should dissuade even the most zealous fans from being much interested in cloning.

One can imagine far more grotesque scenarios than cloning humans that could spin out of the ever-accelerating advances in molecular biology. For example, someone might someday attempt to use a combination of gene transfer and cloning to create novel primates with some "human" capabilities. This dark vision recalls *The Island of Dr. Moreau* by H. G. Wells. The classic science fiction tale recounts the accomplishments of a brilliant physician who, self-banished to a remote Pacific Island, has discovered how to circumvent immunological and genetic barriers in order to create "beast men." Such an act would contradict the teachings of virtually every religion and contravene virtually all ethical systems, as well as violating existing regulations and law. No nation claiming membership in the moral community would permit it. But what if research discovered that the transfer of a particular animal gene into human embryos destined to have a particular disease was the only available cure? Would it be unethical (assuming fully informed consent by the potential parents) to perform the gene transfer? Would the treated fetus be any less human?

What is the most likely scenario by which the first human clone born with a genome copied from an adult will arrive in this world? It is possible that a vainglorious reproductive biologist will clone himself just for the sake of being first, but I think it far more likely that human cloning will occur first in a setting that will be highly likely to evoke our sympathy. For example, imagine a 42-year-old woman who has just lost her only child (a six-week-old infant) and her husband in an auto accident. She arranges for a biopsy of the newly dead child and persuades a friend who is an infertility specialist to transfer a nucleus from the child's cell into one of her own eggs, and then to place it in her uterus. Since she did not witness the infant she lost grow to adulthood, the woman would not watch the cloned child's life unfold with any preconceptions. Who would condemn this act? How would it harm humanity?

How will we respond to the arrival of the first human clone? Will it depend on how we learn? If this news story of the decade is presented as an exclusive television interview or Internet webcast featuring an adoring birth mother and a beautiful infant, is not the clone likely to take on permanent celebrity status (with yearly photo stories in *People* magazine)? I think curiosity and fascination are far more likely to be the dominant responses than is moral revulsion. When the event occurs (and it will), perhaps the society will ultimately regard it as a choice made by a single person (or a couple) for intensely private reasons. If we condemn their act, do we not threaten the general principle of procreative autonomy? Do not most of us passionately believe that marriage and procreation are activities into which the state should rarely, if ever, interfere? Since the world will in no tangible way be harmed by the unusual, perhaps bizarre, choice of a few people to have a child through cloning, then on what basis should we forbid it?

I think that the most powerful argument against human cloning is that there is too great a risk that the child created by cloning will suffer—perhaps severely—when he or she grasps the nature of his origins. It is certainly possible that a human clone would carry no more (and perhaps less) emotional baggage than does a child adopted across racial groups, a child adopted into a gay household, or a child who began life as a donated frozen embryo. Such children are welcomed into our society every day. On balance, it seems the wiser course to err on the side of protecting the potential child and to forbid human cloning for any reason. Of course, forbidding it will not prevent human cloning. It will merely result in a world where it is practiced infrequently and clandestinely. Of course, no matter what the rules, we will still have to anticipate how to respond to the first human clones. The only morally permissible response is to welcome them into the human family.

STEM CELLS

If creating humans by cloning is absolutely immoral, is it ever permissible to clone human *cells* that are totipotent, that is, capable of developing into a human, for other purposes? In 1999, hard upon the world-changing discovery that it was possible to isolate cells from human embryos that have totipotency and grow and store them, this became a dominant bioethical issue that reverberated through Congress and the White House.

Since 1981, scientists have been able to isolate and manipulate mouse embryonic stem cells. This provided the technological foundation upon which transgenic animals were created. In particular, transgenic mice have become one of the most valuable of all tools to use in understanding the role of genes in development and disease.

During the 1980s and 1990s, the conventional wisdom was that human embryonic cells lost totipotency very early and permanently. At the end of 1998, however, two privately funded research laboratories reported convincing evidence that they had isolated and cultured human embryonic stem cells and prevented them from becoming highly differentiated. In 1999 more than a dozen research papers were published extending this

research. Even though it is in its infancy, the work thus far signals a revolution in medical therapy. In the next few years we could see the emergence
of *cell transplant therapy* in which cultured stem cells are persuaded to differentiate into the proper sort of brain cell to be transplanted into persons
with Parkinson disease or to become insulin cells for transplant into dia-
betic patients. In 20 years or so, the cells, mounted on tissue scaffolds, may
become the seeds from which much-needed organs are grown. One can
imagine a world with a virtually unlimited supply of organs available for
those in need.

Reports of the discovery of human stem cells were instantly entangled
in the abortion debate. In the United States and many other nations, it is
illegal to use federal funds to perform research that destroys a human embryo. Yet, by far the best source of these extraordinarily valuable stem cells
are frozen human embryos stored by infertile couples who no longer wish
to attempt pregnancy. The use of tissue obtained from spontaneous abortions is much less attractive to scientists because of the obvious problem
of timing, access to the tissue, risks of infection, and other concerns. The
use of tissue from planned abortions is an alternative possibility, but it has
met with predictably strong opposition.

In 1999 the U.K. imposed a one-year moratorium on experiments
with human stem cells to permit public discourse about the ethical issues.
France's highest court advised lifting that country's ban on human embryo
research. In the United States, the National Bioethics Advisory Commission recommended to President Clinton to permit federal funding of stem
cell research. Late in 1999, the NIH issued guidelines that permitted funding of stem cell research but forbade expenditures to derive the needed cell
lines from human embryos, thus leaving that task to the private sector. In
the halls of Congress, groups lobbying for fetal rights have, somewhat surprisingly, been effectively countered by advocacy groups representing millions of people whose only hope for cure may be stem cell therapy.

The ethical debate over the use of frozen human embryos to serve a
highly desirable line of research that could save so many lives recalls the
public reaction to the daring act a few years ago by a California couple.
When Abe and Mary Ayola, a couple in Walnut, California, learned that
their teenage daughter had a fatal form of leukemia for which the only
possible cure was a bone marrow transplant, they devoted two years of intense effort to finding a donor. To no avail. Already in their early 40s and

feeling that time was running out for them and their daughter, they intentionally conceived a child, hoping that they would win the 1 in 4 bet that the baby would be an immunologically compatible bone marrow donor for their daughter. News of their action evoked sharp criticism from bioethicists who accused them of turning the fetus into a therapeutic object, to which Mrs. Ayola responded, "Our baby is going to have more love than she probably can put up with." The infant did turn out to be an ideal genetic match, and bone marrow taken from her appears to have saved her older sister's life.

Ultimately, the most beneficial use of cloning human cells will probably be the generation of autologous tissue. Perhaps before 2050, we will be able to create our own private organ banks. Some parents are already paying to have cord blood taken from newborn babies and stored. This tissue is rich in stem cells that can be frozen for decades and used if a bone marrow transplant is needed; for example, to treat leukemia. One can imagine a day, decades hence, when cells from an individual will be on call as needed to produce specialized tissues and organs. For example, stem cells might be removed from cord blood and used to create cell lines to treat Parkinson disease and other neurological disorders or grow new organs to replace failing parts. Because this technique does not depend on the creation of embryos, it is not likely to encounter much moral opposition. In the next few years, however, the key research will be done with frozen human embryos. The moral debate will not be resolved, but the political struggle will end in funding the research.

Francis Galton, age 66. *(Reprinted, with permission, from Pearson 1924.)*

Eugenics
Can We Improve the Gene Pool?

In 1949, when she was 15 years old, Betsy Stark, a poor girl who lived in a small town near Lynchburg, Virginia, and who had dropped out of school a year earlier, got pregnant. Shortly after she had the baby, whom she placed for adoption, Betsy was committed (at her father's request) to a large state institution for the mentally retarded in Lynchburg. As part of the admission process, a staff psychologist gave her an IQ test on which she scored 72, considered to be at the border between low-normal intelligence and mental retardation. A few weeks later, after conducting a routine physical exam, a surgeon at the state facility told Betsy that she needed to have her appendix out. Just two weeks after the operation, with no more explanation than she had been given for her admission, Betsy was discharged from the Lynchburg Colony. Fifteen years later, then married and working in a local factory, she learned that the infertility, with which she and her husband had struggled for several years, was due to the fact that the surgeon who had taken out her appendix had, without her knowledge or consent, tied off her fallopian tubes.

Betsy is one of several thousand women who were sterilized pursuant to a 1924 Virginia law designed to protect society from the then widely held suspicion that mentally retarded women were unusually fecund and, unless sterilized, would have large broods of retarded children. Those who lobbied for the law reasoned that these children would almost certainly become wards of the state who would as adults perpetuate and expand the welfare cycle. The Virginia sterilization law earned its niche in history in 1927 when Supreme Court Justice Oliver Wendell Holmes, Jr. wrote the opinion in *Buck vs. Bell* upholding it's constitutionality. He penned the now infamous line, "Three generations of imbeciles are enough." Although they were both sterilized pursuant to the same law, Betsy's story has a decidedly different ending from that of Carrie Buck, the woman who was the subject of the case about which Holmes wrote his opinion.

In the late 1960s, Betsy and hundreds of other women who had been sterilized without their knowledge or consent joined a class action lawsuit against the state of Virginia for deprivation of their civil rights. Although a federal trial court eventually denied most of their claims because the sterilizations had been carried out under a law the constitutionality of which had been upheld by the U.S. Supreme Court, it did order state officials to inform all women who had been surreptitiously sterilized about the true nature of the surgery that had been forced upon them. In 1972, two years before that opinion was published, the governor of Virginia offered a public apology to the women and ordered that no eugenic sterilizations be performed on any persons institutionalized in a state facility. The Virginia class action lawsuit marks the end of an era of state-supported sterilization programs that lasted for more than 60 years in the United States.

EUGENICS: A BRIEF HISTORY

The word eugenics (from the Greek for "good" and "birth") was coined by Francis Galton in 1883 nearly 20 years after he began studying the heritability of talent in British families. Interest in eugenic ideas was even then (well before the discovery of Mendelian laws) easily discernible in England, the United States, and Germany. One important early event in the United States grew out of work by Richard Dugdale, a British emigrant who lived in New York and was an advocate of social reform. In 1875, New York officials asked Dugdale to serve on a committee to inspect the quality of the treatment of inmates in state prisons. While inspecting a prison in upstate New York, Dugdale became fascinated by the number of people in one extended family who were incarcerated. After studying the family for nearly a year, he began to lecture and write about it. In 1877 his book, *The Jukes* (a pseudonym for the family's name), generated immense public interest in the societal costs of caring for the mentally retarded, mentally ill, paupers, alcoholics, and disabled persons.

His history of a family that had over several generations consumed a huge amount of what we would today call welfare dollars, created a new literary genre. Over the next 40 years, more than a dozen book-length, often lurid, accounts of mostly large, poor, rural families, many of whose members were allegedly feebleminded (as the mildly mentally retarded were then called) and who had large broods of children for whom they could not properly care were published. During the 1920s and 1930s, some discussion of *The Jukes,* and *The Kallikaks,* a New Jersey family studied by the eminent psychologist, H.H. Goddard, who worked at the Vineland Training School, was a standard part of the high school biology curriculum, offered as proof of the hereditary nature of feeblemindedness.

Between 1900 and 1910, eugenics attracted huge public interest. The *Reader's Guide to Periodical Literature* for 1910 ranks articles about eugenics as among the most frequently published in the nation's popular magazines. This interest was generated in part by the severe social pressures created by wave after wave of new immigrants, especially as the tide began to flow from southeastern Europe and Russia. Many Americans (who were, of course, themselves descendants of immigrants) were easily convinced by policymakers that new arrivals were of a much weaker stock in the human orchard. First among many ardent eugenicists was President Teddy Roosevelt, who unabashedly urged healthy, young Americans of good stock (presumably of northern European descent) to marry and have large families. This was the only solution, he thought, to avoid the "race-suicide" that would result if swarthy immigrants diluted the existing gene pool.

The influence of eugenic thinking in the United States from the last quarter of the 19th century through the 1960s is most obvious in the language of our restrictive immigration laws. Beginning with the Chinese Exclusion Acts of 1877, federal immigration laws were rewritten time and again to limit entry of those deemed racially (genetically) undesirable. The apogee came in 1924 with enactment of a quota system that gave much greater preference to northern Europeans than to others, a system that was not substantially altered until 1968. The other major expression of our society's interest in "negative eugenics" (programs to curtail reproduction by persons of weak genetic stock) was the enactment in about 30 states of involuntary sterilization laws like the Virginia statute.

During the 1880s and 1890s, an era in which the states built large institutions to house the mentally ill and mentally retarded, eugenic policy was one of segregation. The literature of the time is replete with discussions of the need to protect retarded women from being "taken advantage of." Officials who ran these facilities strictly enforced rules that eliminated this possibility. But segregation by gender presumed that persons would never leave the institution, a plan that was in conflict with a larger mission. These same officials believed that they could educate many of the feeble-minded well enough to prepare them to hold simple jobs in the community.

The rediscovery of the Mendelian laws of inheritance in 1900 provided a neat mechanism to explain mysterious conditions like mental retardation and insanity. Charles Davenport, a Harvard-trained biologist who founded the now famous research laboratory at Cold Spring Harbor, Long Island, New York, was one prominent scientist who taught that a sizable fraction of persons afflicted with such conditions had either autosomal dominant or autosomal recessive disorders. Enamored with Mendelism (as it was called before the term genetics appeared about 1908), Davenport thought heredity forces could explain just about any condition or behavior. He even postulated that because boys, but not girls, ran off to sea, there must be an X-linked gene driving this desire, a condition that he dubbed thalassophilia (from the Greek for "sea" and "love").

The development of the vasectomy in the United States about 1900 provided an important tool for eugenics. The first involuntary sterilization law was enacted in Indiana in 1907, thanks to the lobbying efforts of a prison doctor named Harvey Sharp, who was the first surgeon to use the vasectomy to sterilize convicted felons. California enacted a similar law in 1909 that focused on mentally ill persons living in state institutions. In less than a decade, state officials working in California hospitals sterilized several thousand people, a most impressive number, given the population of the state at that time. However, except for California, the dozen or so states that enacted sterilization laws before World War I did not implement them aggressively. In the early 1920s, in large part in response to a resurgence of European immigration after the war, there was a dramatic growth in public support of eugenic policies. Driven by the lobbying efforts of groups like the American Eugenics Society, Michigan and Virginia became the first states in the postwar era to enact sterilization laws aimed at retarded persons in state institutions. When the Supreme Court upheld the Virginia statutes in 1927, many states quickly enacted similar laws. Between 1927 and 1939 more than 40,000 persons were sterilized pursuant to these laws.

During the 1930s, interest in eugenics was widespread throughout Europe. England never enacted a sterilization law, but a dozen or more European states did. Nazi Germany eclipsed all others when in 1934 it created a system of "hereditary health courts" which were charged with hearing petitions from local officials that certain citizens were unfit for parenthood and should be sterilized. During 1934 alone, the special Nazi health courts ordered the sterilization of more than 80,000 persons, a number that probably approximates the combined total of all involuntary sterilizations performed in all other nations during the years in which eugenic thinking held sway. Eugenics took on insane dimensions in Nazi Germany, and more than 1,000,000 sterilizations may have been performed there.

The notion that negative eugenic policies could be rationally used to improve the human gene pool is quixotic at best, madness at worst. Such policies teeter on a child's block tower of assumptions. Is there a real causal relationship between genetic endowment (or lack thereof) and a particular condition or behavior? How does a society decide which of the thousands of genetically influenced human conditions should be diminished? To what extent will decisions reflect transient political assumptions? Do we know enough about genetics to argue that any gene is "bad" or "good?" No one has the answers to these questions.

Nearly 20 years ago, I visited a half-dozen old state institutions for the mentally retarded, traveling to places like Faribault, Minnesota, and Lynchburg, Virginia, to poke through their dusty archives and to read the medical records of persons now long dead. I sought to learn how they implemented eugenic sterilization laws and to unearth the private opinions of some of the administrators and doctors who seemed eager to carry them out. The old medical records speak volumes. The institutional physicians clearly believed that both the young, mildly retarded women who were by 1930 the most usual target of sterilization and the society as a whole would benefit from these operations. In several states I found evidence that consulting surgeons had routinely performed an extraordinarily high number of "appendectomies." It is simply not medically possible that 20% or more of the young women admitted to state institutions for the mentally retarded each year needed appendectomies. Most, if not all, of these operations were almost certainly surreptitious tubal ligations. No one will ever know how many hundreds (probably thousands) of women in the United States were sterilized while in state institutions (which often had medical facilities on grounds) without any record even being made of it.

Six decades have elapsed since the heyday of these state programs intended to protect the gene pool by sterilizing retarded persons. We now know that the origins of mental retardation are manifold and that sterilization of a few thousand women and men could not have a discernible impact on the gene pool or measurably reduce the number of people in the subsequent generation with mental retardation. In fact, contrary to the popular prejudice of earlier times, relatively few persons with mental retardation have children, and most persons with mental retardation are born to couples who do not suspect their risk. Furthermore, many cases of mild mental retardation are probably caused, or at least exacerbated by, events in pregnancy and early childhood, especially forces linked to poverty, as much as by genetic factors.

In the United States and elsewhere, the fear of our eugenic legacy lingers. In the 1970s, some in the black community genuinely viewed well intentioned, but poorly drafted, state laws that mandated testing for sickle cell anemia to be a thinly disguised, racially motivated effort to dissuade gene carriers from reproducing. Today, many disabled persons are alarmed that prenatal diagnostic testing is sharply reducing the births of children with spina bifida. In England and Wales, prenatal screening to identify fetuses at risk for such neural tube defects, coupled with selective pregnancy termination, has led to more than a 90% reduction in the births of affected persons in just two decades. Once among the most common birth defects in England, spina bifida is rapidly becoming rare. The same trend is discernible in the United States. Is this a public health achievement that we should applaud or castigate?

Elsewhere in the world, people have harnessed medical technologies to serve cultural values that most Americans and Europeans find abhorrent. Astonishingly, researchers at the United Nations argue that there may be as many as 100 million women missing from the world population. This is because across Asia, parents favor the survival of girls over boys. In China, which for 20 years has officially permitted only one child per family (a policy now beginning to loosen slightly), parents often deliberately neglect illness in female infants. In India and China, despite laws that criminalize the activity, ultrasonography is still easily available to sex fetuses, and among the well-to-do, the use of selective abortion to control gender is common. In India, a 1991 census found only 92.9 females for every 100 males; in China, a 1990 census found 93.8 females per 100 males. These numbers are dramatically out of line with expected ratios and can only be explained by selective abortion or by behaviors that reduce the survival chances of young girls.

Even if one could distinguish gender discrimination from eugenics, China has embraced other social policies that make it difficult to dismiss those who call up the ghost of eugenics past. On June 1, 1995, China implemented a new law on Maternal and Infant Health Care. Although the law has many laudable elements (such as a provision that forbids using ultrasound or chromosome studies to diagnose fetal sex), many in the West are troubled by the provisions that *require* couples known to be at risk for having children with serious genetic disorders to receive and follow medical counseling, and to agree to long-acting contraception, sterilization, or prenatal diagnosis and selective abortion if recommended by a physician.

The most troublesome language in the Chinese law is Article X, which states that during a prenatal exam, the doctor "shall explain and give medical advice to both the male and female who have been diagnosed with certain genetic diseases of a serious nature which is considered inappropriate for child-bearing from a medical point of view." Another provision, Article XVI, of the law seems to empower the doctor to control the childbearing behavior of at-risk couples who seem recalcitrant to advice.

Geneticists in several nations, notably Great Britain, have urged the Chinese government to rescind eugenic features of the law. At an International Congress of Genetics held in Rio de Janeiro in 1996, Newton Morton, one of the world's foremost population geneticists, spearheaded an effort to urge geneticists to ask the Chinese government to rescind the law, and to call for a moratorium on similar laws in other nations. Uncomfortable with censoring a particular nation, the American Society of Human Genetics instead appointed a committee to draft a general policy statement on eugenics and reproductive freedom. I chaired that committee, which spent two years navigating through murky political seas. The ASHG ultimately adopted a statement based on the committee's work, a document that strongly opposes any government effort to compromise procreative autonomy.

The discontent among Western geneticists over the Chinese eugenics law is not shared by their Chinese colleagues. A 1996 survey of 402 geneticists in 30 provinces and autonomous regions of China found that 95% agreed that "people at high risk for serious disorders should not have children unless they use prenatal diagnosis and selective abortion," and 90% agreed that "an important goal of genetic counseling is to reduce the number of deleterious genes in the population." As one Chinese scientist put it, "Please remember that we are already caring for 60,000,000 disabled people," a number about equivalent to the population of France.

THE NEW NEGATIVE EUGENICS

In the United States and Europe, a new and more subtle form of negative eugenics may be emerging. It is technologically enabled, physician supported, and widely embraced by consumers. Simply put, a significant fraction of women, when offered testing that could lead them to avoid the birth of children with serious genetic or congenital disorders, choose to be tested and to abort affected fetuses. Amniocentesis and fetal chromosome analysis became widely available in the 1970s. About 300,000 pregnant women (most 35 and older) in the United States now undergo this test each year. About 1,000 will discover that they are carrying a fetus with Down syndrome, and three-quarters of them will obtain an abortion. About 3 out of 4 of all pregnant women (that is about 3,000,000 women) will, regardless of age, undergo a screening test to see if the fetus may have spina bifida. About 3,000 will eventually discover that they are carrying an affected fetus, and about three-quarters of them will abort.

For a few severe childhood genetic disorders that are relatively common in specific populations, such as Tay-Sachs disease among Ashkenazi Jews, carrier testing to alert individuals about reproductive risk has become the norm. In the United States, carrier testing, prenatal diagnosis, and selective abortion have reduced the births of children with Tay-Sachs disease by more than 95% in 25 years. As the diagnoses of genetic disorders are made ever earlier, physicians are able to alert parents about high recurrence risks in subsequent pregnancies. For that reason, we can expect that there will over time be dramatically fewer families into which a second affected child with a serious genetic disorder is born.

The point is simple and the trend is clear. In advanced societies that have embraced small families, widespread use of these testing technologies by women is causing a sharp reduction in births of children with some of the more common birth defects. Although such tests will greatly affect individual families, they will have no discernible impact on welfare budgets and none on the gene pool. The reason for the latter is that virtually all of these "prevented lives" would have been people who would not have become parents.

Interest in using prenatal testing to avoid the births of children with serious disorders is growing just as we enter an era in which DNA-based testing will provide a greatly expanded array of tests. Once the cost of testing becomes modest, which it will, there are several ways in which the tests will be used, including (1) population-based newborn screening to identify children at risk for disorders that manifest during infancy or childhood, (2) carrier testing of adults to warn people about risks (depending on their choice of spouse) of bearing children with autosomal recessive disorders, or to warn women about their risk of bearing children with X-linked disorders, and (3) prenatal diagnosis. Newborn screening is already well established and will naturally expand to include tests that benefit children. Carrier testing is also well established for three disorders (Tay-Sachs disease, sickle cell anemia, and the thalassemias) and is likely to expand sharply to include testing for cystic fibrosis, Fragile X, and several other conditions in the next few years. The future of DNA-based testing in prenatal diagnosis is uncertain. I expect it to develop slowly, largely because we know so little about the pattern of interactions of genes and environmental factors that lead to many birth defects. But develop it will. Twenty years hence a woman will have the option of learning vastly more information about her fetus than is currently possible. Just as the prenatal ultrasound has become a routine part of prenatal care, so will DNA screening.

One prediction does seem certain. Women in large numbers will continue to use genetic (and other) tests to avoid the birth of children with serious disorders. Modern eugenic practices do not derive from, nor do they depend on, governmental funds or policies. If there is a new eugenics in the land, it is certainly not driven by state law. If anything, in the many states in which the majority of people favor restricting access to abortion services, it can be difficult for indigent women to obtain prenatal tests.

It is time for us to confront a discomforting fact. Vast increases in our understanding of the role genes play in health and disease and in our ability to acquire and process such information about the risks faced by couples or about fetuses conceived by them will have a frightening impact on how we think about human reproduction. Who will decide what information is of sufficient clinical value to offer to couples? Who is to be charged with deciding what constitutes a serious genetic condition? Who will act as the gatekeepers to block consumers from gaining access to information they desire? Are we entering an era in which even relatively minor genetic or congenital conditions will be viewed as disabilities? The process by which we resolve these issues may make the abortion debate seem mild.

POSITIVE EUGENICS

Frances Galton's earliest interest was in *positive* eugenics, programs to improve the mean genetic endowment of the future population. The goal of positive eugenics has always been to think of ways to increase the reproductive contribution of those individuals thought to be most fit. For 50 years, from Teddy Roosevelt to the Nazi Lebensborn program, this amounted to little more than periodic exhortations to healthy young couples to have big families. In the United States, the popularity of such ideas peaked in the 1930s. This was a decade in which county fairs sponsored "fitter family" contests, awarding blue ribbons for the best human stock just as they did for the best cows and pigs. It was an era that saw the emergence of nonprofit organizations with names like The Human Betterment Foundation (California) and the Eugenic Babies Foundation (Kansas).

Nobel Prize-winning geneticist, H. J. Muller (who first observed that radiation caused mutations), was for a time an ardent positive eugenicist. Muller was among the first to propose the use of sperm donations by eminent men (presumably to be used by enlightened couples) to differentially increase their genetic contribution to the everchanging gene pool. Positive eugenics reached its apotheosis among the Nazis, who encouraged young Aryan couples to marry, and even recruited young women to become pregnant by soldiers of presumably good genetic stock whom they barely knew. In the 1970s a few other scientists (most notoriously Alfred Shockley, a Nobel laureate in physics) suggested that the use of donor sperm from highly successful men might be the best available way to secure a superior genetic endowment for a child.

Perhaps the most famous modern proponent of sperm banking for positive eugenics is Robert Klark Graham, a retired lens manufacturer, and millionaire, who in 1980 opened the Repository for Germinal Choice in Escondido, California. The idea was to select and purchase sperm donations from Nobel Prize-winners and offer them to deserving women (whose applications he and his wife evaluated) who wished to have highly intelligent children. Still open and more active than ever, the Repository no longer seeks sperm donors among Nobel laureates (using the sperm of older men is less likely to result in a pregnancy). One of Graham's assistants told me that in recruiting donors they now focus on successful scientists and other professional men in their 30s and 40s who have a tested IQ of at least 135.

More than 200 children in the United States have been conceived with sperm samples selected by women after they have reviewed the data (height, hair color, ethnic origin, IQ, hobbies, musical ability, and physique) provided by the anonymous donors. According to unverified reports from the mothers, all the children (some now approaching adulthood) are very bright. A few have IQs in the 170s, a score to be found in only about 1 in 100,000 randomly tested children. Most of these kids are being reared by unmarried, professional women who have substantial resources to devote to them. Thus, the relative contributions of genes and environment are impossible to untangle.

California Cryobank, one of the largest commercial (there are also many small ones operated by infertility centers) sperm banks operating in the United States, also seems to have embraced Muller's views. When it decided recently to open a branch on the East coast, it chose to set up shop within a stone's throw of Harvard and MIT. The company, which pays its donors $35 per sample and will accept up to three samples a week, claims that only 8% of those who apply to be donors are accepted. This is about the same percentage of applicants that Harvard and MIT accept as freshmen! Although the company does not require IQ test scores of applicants, it is eager to collect sperm from college students who have already demonstrated obvious talent. Of course, evidence of good physical health is also essential. Other baseline requirements? A donor has to be at least 5 feet, 9 inches tall.

In 1999 the idea that one might shop for germ cells was taken to a new level when Ron Harris, a longtime fashion photographer, launched a website featuring the photos of eight beautiful models, all of whom were allegedly willing to sell their eggs for prices ranging up to $150,000. In a *New York Times* interview, Harris claimed that one of his models ("Ron's angels") had already received an offer of $42,000 (of which he would get a

20% commission). Within a day or two of being launched, the website came under heavy scrutiny. Most of the models withdrew from the list of potential egg donors, and the site did not respond to efforts by reporters to bid. It appears to be just another Internet-based hoax. Legitimate Internet auction sites, like eBay, will not permit sale of any human tissue except hair.

Ron's angels may be a fraud, but the question it raises is real enough. Should women be able to auction off their eggs to the highest bidder? A few voluntary egg donation programs are in operation in the United States, all closely tied to legitimate infertility centers. Egg donors are paid a modest amount (between $2,500 and $5,000) which is considered a service fee for the inconvenience they suffer, not a purchase price. Several professional and consumer groups have advocated that the sale of human eggs be outlawed. Yet, even in the most altruistically motivated infertility clinic, couples searching for an egg are given the opportunity to review fact sheets about the donors that include information like height, weight, and complexion. Where should the line be drawn? Is it OK to list the donor's IQ?

It would be foolish to expect that a child born after IVF with an egg donated by a beautiful woman would grow up to be beautiful. However, it would also be naive not to think that the odds of that child being beautiful had been improved substantially because of the DNA in that egg. Perhaps Mr. Harris is right. "This is Darwin's natural selection at its very best," he wrote. "The highest bidder gets youth and beauty."

There may have been a time, more than 20,000 years ago, when the human population was composed of as few as 10,000 hunters and gatherers living in nomadic bands, that a comparatively few people had a fairly large impact on the gene pool. The male leaders of each clan may have sired a significant fraction of the offspring. To the extent that this social dominance reflected a superior (for that particular environment) genetic endowment, they may have acted as a vector of positive eugenics. This is a highly speculative idea that has no relevance to human culture today.

The idea that a complex modern society can enhance its genetic endowment by encouraging differentially high fertility by certain citizens is little more than an adolescent fantasy, the kind of thinking one might find among people who take *Star Trek* a little too seriously. Yet, research suggests that a significant fraction (40–60%) of the variance in intelligence is attributable to genes. If this is so, then someday in the distant future, we will identify and elucidate the role played by those genes in cognition. Will we ever be able to genetically enhance the intellectual potential, physical prowess, musical aptitude, beauty, or sociability of our children? Perhaps. But that will be many generations from now in a world so vastly different from the one we now know that it cannot be imagined.

Bibliography

CHAPTER 1

ABRAHAM LINCOLN: DID HE HAVE MARFAN SYNDROME?

Pyeritz R. and Conant J. 1989. *The Marfan syndrome*, 3rd ed. National Marfan Foundation, Port Washington, New York.

Sandburg C. 1960. *Abraham Lincoln*. Dell, New York.

Advisory Statement by the Panel on DNA Testing of Abraham Lincoln's Tissue. McKusick V., Chairman. Spring 1991. Caduceus National Museum of Health and Medicine, Washington, D.C., pp. 41–47.

Dietz, H., Cutting G., Pyeritz R., Maslen C., Sakai L., Corson G., Puffenberger E., Hamosh A., Nanthakumar E., Curriston S., Stetten G., Meyers D., and Francomani C. 1991. Marfan syndrome caused by a recurrent missense mutation in the fibrillin gene. *Nature* 352:337–339.

Leary W. February 10, 1991. Scientists seek Lincoln DNA to clone for a medical study. *The New York Times*, p.1.

McKusick V. 1991. The defect in Marfan syndrome. *Nature* 352:279–281.

Pyeritz R. and McKusick V. 1979. The marfan syndrome: Diagnosis and management. *N. Engl. J. Med.* 300:772–776.

Abraham Lincoln Association at www.alincolnassoc.com

National Marfan Foundation at www.marfan.org

CHAPTER 2

KINGS AND QUEENS: GENETIC DISEASES IN ROYAL FAMILIES

Ayling S. 1972. *George the Third*. Knopf, New York.

Ginsburg, D., 1996. Hemophilias and other disorders of hemostasis. In *Principles and practice of medical genetics*, 3rd ed., vol II (ed. Rimoin D., Connor J. and Pyeritz, R.) Churchill Livingstone, New York, pp. 1651–1675.

McGovern, M. Anderson K., Astrin K., and Desnick R. 1996. Inherited Porphyrias. In *Principles and practice of medical genetics*, 3rd ed., vol. II (ed. Rimoin D., Connor J., and Pyeritz R.) Churchill Livingstone, New York, pp. 2009–2036.

Potts M. and Potts W. 1995. *Queen Victoria's gene*. Alan Sutton, Stroud, United Kingdom.

Stern C. 1972. *Principles of human genetics*, 3rd ed. W.H. Freeman, San Francisco, pp. 146–59.

Witts L. 1972. Porphyria and George 3d. *Br. Med. J.* 277:479–480.

The British Monarchy at www.royal.gov.uk/history/george.htm

National Hemophilia Foundation at www.hemophilia.org

Chapter 3

Toulouse-Lautrec: An Artist despite His Genes

Ackroyd C. 1989. *Toulouse-Lautrec.* Chartwell Books, London.

Frey J. 1994. *Toulouse-Lautrec: A life.* Viking, New York.

Angier N. June 6, 1995. What ailed Toulouse-Lautrec? Scientists zero in on a key gene. *The New York Times*, p. C3.

Bittles A. and Neel J. 1994. The costs of human inbreeding and their implications for variations at the DNA level. *Nat. Genet.* 8:117–121.

Castleman R., and Wittrock W. 1985. *Henri de Toulouse-Lautrec.* Museum of Modern Art, New York.

Frey, J. What dwarfed Toulouse-Lautrec? *Nat. Genet.* 10:128–130.

Gelb B., Edelson J., and Desnick R. 1995. Linkage of pycnodysostosis to chromosome 1q21 by homozygosity mapping. *Nat. Genet.* 10:235–239.

Maroteaux, P. and Lamy, M. 1965. The malady of Toulouse-Lautrec. *J. Am. Med. Assoc.* 191:715–717.

Henri de Toulouse-Lautrec, Life and Graphic Works at www.sandiegomuseum.org/lautrec/lautrec_bib.html

Chapter 4

Old Bones: DNA and Skeletons

Massie R. 1995. *The Romanovs: The final chapter.* Johnathan Cape, London.

Editorial. 1996. Romanovs find closure in DNA. *Nat. Genet.* 12:339–340.

Gill P., Ivanov P., Klimpton C., Piercy R., Benson N., Tully G., Evett I., Hagelberg E., and Sullivan K. 1994. Identification of the remains of the Romanov family by DNA analysis. *Nat. Genet.* 6:130–135.

Goodman A. September 7, 1999. A furor for Velazquez: His art but also his bones. *The New York Times*, p.C2.

Ivanov P., Wadhams M., Roby R., Holland M., Weedn V., and Parsons T. 1996. Mitochondrial DNA sequence heteroplasmy in the Grand Duke of Russia Georgij Romanov establishes the authenticity of the remains of Tsar Nicholas II. *Nat. Genet.* 12:417–420.

MacIntyre B. July 9, 1995. Jesse James and science in a final showdown. *The Times* (London), p. 1.

Wade N. May 9, 1999. DNA backs a tribe's tradition of early descent from the Jews. *The New York Times*, p. C2.

Chapter 5

DNA Detectives: The New DNA Evidence

Ballantyne J., Sensabaugh G., and Witkowski J., eds. 1989. *DNA technology and forensic science* (Banbury Report 32). Cold Spring Harbor Laboratory Press, Cold Spring Harbor, New York.

Billings P., ed. 1992. *DNA on trial: Genetic identification and criminal justice.* Cold Spring Harbor Laboratory Press, Cold Spring Harbor, New York.

Committee on DNA Technology in Forensic Science. 1992. *DNA technology in forensic science*. National Academy Press, Washington, DC.

U.S. Department of Justice. 1991. *Forensic DNA analysis: Issues,* (NCJ 128567). Office of Justice Programs, Rockville, MD 20850.

Butterfield F. February 5, 1996. New clues in an old murder case: DNA evidence suggests Sheppard did not kill his wife in '54. *The New York Times,* p. B1.

Hagleberg E., Gray I. and Jeffreys A. 1991. Identification of the skeletal remains of a murder victim by DNA analysis. *Nature* 352:427–429.

Menotti-Raymond M., David V., and O'Brien S. 1997. Pet cat hair implicates murder suspect. *Nature* 386:774.

Wade N. May 9, 1999. DNA backs a tribe's tradition of early descent from the Jews. *The New York Times,* p.1.

Chapter 6

Cold Hits: The Rise of DNA Felon Databanks

Wambaugh J. 1989. *The blooding.* William Morrow, New York.

Ballantyne J. 1997. Mass disaster genetics. *Nat. Genet.* 15:329–330.

Dedman B. October 7, 1999. A rape defendant with no identity, but a DNA profile. *The New York Times,* p.1.

Henry L. November, 1993. Evidence at 80 Below. *Minnesota Monthly,* pp. 56–59, 110–114.

McEwen J. 1995. "*State DNA forensic databanking: Legal, ethical, and social policy implications and recommendations for the future.*" Ph.D. thesis, UMI Dissertation Services (UMI Number 9629778). University of Michigan, Ann Arbor.

McEwen J. and Reilly P. 1994. A review of state legislation on DNA forensic databanking. *Am. J. Hum. Genet.* 54:941–958.

Miller W. August 21, 1994. Britain plans DNA crime database. *Boston Globe,* p. B6.

National Commission on the Future of DNA Evidence. September 1999. Post-conviction DNA testing: Recommendations for handling requests. (NCJ 177626.) National Institute of Justice, Washington, D.C.

Nowell P. November 5, 1999. Police association seeks DNA testing of all suspects. *Boston Globe,* p.A24.

Rosenthal N. 1994. Tools of the trade—Recombinant DNA. *N. Engl. J. Med.* 331:315–317.

Chapter 7

Genes and Violence: Do Mutations Cause Crime?

Mednick S. and Christiansen K., eds. 1977. *Biosocial bases of criminal behavior* Gardner Press, New York.

Brunner H., Nelen M., Breakefield X., Ropers H., and Van Oost B. 1993. Abnormal behavior associated with a point mutation in the structural gene for monomine oxidase A. *Science* 262:578–583.

Coffey M. 1993. The genetic defense: excuse or explanation? *William and Mary Law Rev.* 35:353–398.

Gibbs W. March 1995. Seeking the criminal element. *Sci. Am.* p.100–107.

Hook E. 1973. Behavioral implications of the human XYY genotype. *Science* 179:139–145.

Jacobs P., Brinton M., Melville M., and Norton P. 1965. Aggressive behavior, mental subnormality and the XYY male. *Nature* 208:1351–1353.

Jones K.L. 1988. *Smith's recognizable patterns of human malformation,* 4th Edition, p. 65. W.B. Saunders Company, New York.

Mobley v. Georgia 265 Ga. 292, 455 S.E. 2d 81 (1995).

Restak R. July/August 1992. See no evil: The neurological defense would blame violence on the damaged brain. *The Sciences,* p. 16–21.

Wilson J.Q. and Herrnstein R.J. 1985. *Crime & Human Nature.* Simon and Schuster, New York.

Witkin P. and Mednick C. 1976. Criminality in XYY and XXY men. *Science* 193:54–55.

Wright R. February, 1995. The biology of violence. *The New Yorker,* pp. 67–77.

CHAPTER 8

WRONGFUL BIRTH: WHAT SHOULD THE DOCTOR KNOW?

Biesecker B. 1999. Genetic Counseling: practice of genetic counseling. In *Encyclopedia of bioethics,* Vol. 2 (ed. Reich, W.) MacMillan Publishing, New York, pp. 923–927.

Brock D.J.H., Rodeck C.H., and Ferguson-Smith M.A., eds. 1992. *Prenatal diagnosis and screening,* p. 120. Churchill Livingstone, Edinburgh.

Frankel M. and Teich A., eds. 1994. *The genetic frontier: Ethics, law, and policy,* American Association for the Advancement of Science, Washington, DC.

Capron A. 1973. Informed decision making in genetic counseling: A dissent to the wrongful life debate. *Indiana Law J.* 48:581–624.

Fragile X Advocate. A quarterly journal published by Avanta Media Corporation, P.O. Box 17023, Chapel Hill, NC 27516–1702.

Fragile X Mental Retardation 3 May/June, 1988. Proceedings of the Third International Workshop on Fragile X and X-linked Mental Retardation (ed. Opitz, J.). *Am. J. Med. Gen.* 30(1/2).

Pratt v. University of Minnesota Affiliated Hospitals and Clinics 414 N.W. 2d 399 (Minn. 1987).

Vaccaro v. Milunsky 406 Mass. 777 (1990).

Wertz D. and Fletcher J. 1991. Privacy and disclosure in medical genetics examined in an ethics of care. *Bioethics* 5:212–232.

CHAPTER 9

MENTAL ILLNESS: HOW MUCH IS GENETIC?

Tsuang M. and Faraone S. 1990. *The genetics of mood disorders.* Johns Hopkins University Press, Baltimore.

Vandenberg S., Singer S. and Pauls, D. 1986. *The heredity of behavior disorders in adults and children.* Plenum Press, New York.

Winokur G.1991. *Mania and depression: A classification of syndrome and disease.* Johns Hopkins University, Baltimore.

Egeland, J. Gerhard D., Pauls D., Sussex J., Kidd K., Allen C., Hostette A., and Housman D. 1987. Bipolar affective disorders linked to DNA markers on chromosome 11. *Nature* 325:783–787.

Lander E. 1988. Splitting schizophrenia. *Nature* 336:105–106.

Pauls D. 1993. Behavioral disorders: Lessons in linkage. *Nat. Genet.* 3:45.

Risch N. and Botstein D. 1996. A manic depressive history. *Nat. Genet.* 12:351–353.

Schmeck H. November 7, 1989. Scientists now doubt they found faulty gene linked to mental illness. *The New York Times,* p.C3.

Sherrington R., Brynjolfsson J., Petursson H., Potter M., Dudleston K., Barraclough B., Wasmuth J., Dobbs M., and Gurling H. 1988. Localization of a susceptibility locus for schizophrenia on chromosome 5. *Nature* 336:164–167.

CHAPTER 10

PERSONALITY: WERE WE BORN THIS WAY?

Bouchard T. 1994. Genes, environment, and personality. *Science* 264:1700–1701.

Bouchard T., Lykken D., McGue M., Segal N., and Tellegen A. 1990. Sources of human psychological differences: The Minnesota study of twins reared apart. *Science* 250:223–228.

Carrington M., Nelson, G., Martin, M., Kissner T., Vlahov D., Goedert J. J., Kaslow R., Buchbinder S., Hoots K., and O'Brian S.J. 1999. HLA and HIV-1: Heterozygote advantage and B*35-Cw*04 disadvantage. *Science* 283:1748–1752.

Cloninger R., Adolffson R., and Svrakic N. 1996. Mapping genes for human personality. *Nat. Genet.* 12:3–4.

Ebstein R. Novick O., Umansky R., Priel B., Osher Y., Blaine D., Bennett E., Nemanov M., and Belmaker R. 1996. Dopamine D4 receptor (D4DR) exon III polymorphism associated with the human personality trait of novelty seeking. *Nat. Genet.* 12:78–80.

Editorial. March 1999. What we learn from twins. *The Economist.* pp.74–81.

Eiberg H., Berendt I., and Mohr J. 1995. Assignment of dominant inherited nocturnal enuresis (ENUR1) to chromosome 13q. *Nat Genet.* 10:354–356.

Hamer D. 1996. The heritability of happiness. *Nat. Genet.* 14:125–126.

Plomin R. 1990. The role of inheritance in behavior. *Science* 248:183–188.

Plomin R., Owen M., and McGuffin P. 1994. The genetic basis of complex human behaviors. *Science* 264:1733–1739.

Skuse D., James R., Bishop D., Coppin B., Dalton P., Asmodt-Leeper G., Bacarese-Hamilton M., Creswell C., McGurk R., and Jacobs P.A. 1993. Evidence from Turner's syndrome of an imprinted X-linked locus affecting cognitive function. *Nature* 387:705–708.

CHAPTER 11

TALENT: NATURE OR NURTURE?

Herrnstein R. and Murray C. 1994. *The bell curve.* Free Press, New York.

Sokal, Michael ed. 1990. *Psychological testing and American society: 1890-1930.* Rutgers University Press, New Brunswick, New Jersey.

Terman L. And Oden M., eds. 1925. *Genetic studies of genius, Volume I. Mental and physical traits of a thousand gifted children.* Stanford University Press, Stanford, California.

Adamson L. Aug. 17, 1997. When creativity runs in the family. *The New York Times,* Sec 2, p.1.

Blakeslee S. Nov. 20, 1990. Perfect pitch: The key may lie in the genes. *The New York Times.* p. C11.

Browne M. June 6, 1997. Genes a lifelong factor in intelligence. *The New York Times,* p.A16.

Conrad P. June 6, 1993. Natasha Richardson and the Redgrave Dynasty. *The New York Times Magazine,* p.40, 52–55.

Glanz J. November 5, 1999. Study links perfect pitch to tonal language. *The New York Times,* p.1.

Goleman D. March 7, 1995. 75 years later, study still tracking geniuses. *The New York Times,* p. C1.

Gould S. 1990. *The mismeasure of man.* Norton, New York

McClearn G., Johansson B., Berg S., Pedersen N., Ahern F., Petrell S., and Plomin R. 1997. Substantial genetic influence on cognitive abilities in twins 80 or more years old. *Science* 276:1560–1563.

Plomin R. 1999. Genetics and general cognitive ability. *Nature* (Suppl.) 402:C25–29.

Wade N. Sept. 7, 1999. Of smart mice and an even smarter man. *The New York Times,* p.1F.

Wright L. Aug. 7, 1995. Double mystery. *The New Yorker,* pp. 45, 53–59.

University of California, San Francisco, The Perfect Pitch at www.perfectpitch.org

Chapter 12

Gay Genes: What's the Evidence?

Burr C. 1996. *A separate creation: The search for the biological origins of sexual orientation.* Hyperion, New York.

Hamer D. and Copeland, P. 1994. *The science of desire: The search for the gay gene and the biology of behavior.* Simon & Schuster, New York.

Bailey J. 1995. Sexual orientatation revolution. *Nat. Genet.* 11:33–34.

Bailey J. and Pillard R. 1991. A genetic study of male sexual orientation. *Arch. Gen. Psychol.* 48:1089–1096.

Barinaga M. 1991. Is homosexuality biological? *Science* 256:956–957.

Burr C. March 1993. Homosexuality and biology. *Atlantic Monthly,* p.47.

Goode E. April 23, 1999. Study questions gene influence on male homosexuality. *The New York Times,* p. A19.

Hamer D., Hu S., Magnuson, V., Hu N., and Pattatucci M. 1993. A linkage between DNA markers on the X chromosome and male sexual orientation. *Science* 261:321–327.

Hubbard R. Aug 2, 1993. False genetic markers. *The New York Times,* p. A15.

Rice, G., Anderson, C., Risch, N., and Ebers, G. 1999. Male homosexuality: Absence of linkage to microsatellite markers at Xq28. *Science* 284:665–667.

CHAPTER 13

GENETICALLY MODIFIED ORGANISMS: THE NEXT GREEN REVOLUTION?

Bowers J. and Meredith C. 1997. The parentage of a classic wine grape. Cabernet sauvignon. *Nat. Genet.* 16:84–87.

Bowers J., Boursiquot J-M., This P., Chu K., Johansson H., and Meredith, C. 1999. Historical genetics: The parentage of chardonnay, gamay, and other wine grapes of northeastern France. *Science* 285:1562–1565.

Estruch J., Carozzi N., Desai N., Duck N., Warren G., and Koziel M. 1997. Transgenic plants: An emerging approach to pest control. Bio/Technology 15:137–141.

Firth P. May 1999. Leaving a bad taste. *Sci. Am.* 280:34–35.

Luby J. 1997. Elegant nose and powerful body—Who were the parents of such nobility? *Nat. Genet.* 16:4–5.

Plant Biotechnology: Food & Feed. 1999. Special section. *Science* 285:367–393.

Reibstein L. and Beals G. March 10, 1997. A cloned chop? Anyone? *Newsweek,* pp.58–59.

The Royal Society Advisory Committee on Genetically Modified Foods. Sept. 1998 Genetically Modified Plants for Food Use (www.royalsoc.ac.uk/st_po140.htm).

Thompson, P. 1997. Food biotechnology's challenge to cultural integrity and individual consent. *Hastings Center Rep.* 27:35–38.

Yoon C. May 20, 1999, Altered corn may imperil butterfly, researchers say. *The New York Times,* p.1

Yoon C. Nov 3, 1999. Squash with altered genes raises fears of 'superweeds'. *The New York Times,* p. A1.

U.S. Department of State, International Information Program, Global Issues, Biotechnology at www.usia.gov/topical/global/biotech

CHAPTER 14

TRANSGENIC ANIMALS: NEW FOODS AND NEW FACTORIES

Abbott A.1996. Transgenic trials under pressure in Germany. *Nature* 380:94.

A/F Protein, Inc. The material about A/F Protein, Inc., is partly based on remarks by its CEO, Eliot Entis at the 1997 meeting of the Biotechnology Industry Organization in Houston.

Allen S. Aug. 30, 1999. Some aren't hooked on superfish 'revolution' *Boston Globe,* p. A1.

Genzyme Transgenics. The remarks about Genzyme Transgenics are based in part on remarks by its CEO, James Geraghty, at the 1997 meeting of the Biotechnology Industry Organization in Houston, Texas.

Gershon D. 1993. Prospects for growth hormone turn sour. *Nature* 364:565.

Juskevich J.C. and Guyer C.G. 1990. Bovine growth hormone: Human food safety evaluation. *Science* 249:875–884.

Miller H. 1993. Foods of the future: the new biotechnology and FDA regulation. *J. Am. Med. Assoc.* 269:910–912.

O'Donnell P. October 6, 1996. EU concern over genetically modified foods. *Bioworld,* pp.4, 6.

Torrado C., Bastian W., Wisniewski K., and Castells S. 1991. Treatment of children with Down syndrome and growth retardation with recombinant human growth hormone. *J. Pediatr.* 119:478–483.

CHAPTER 15

ENDANGERED SPECIES: NEW GENES BEAT EXTINCTION

Plotkin, M.1993. *Tales of a shaman's apprentice.* Viking, New York.
Wilson E. O. 1992. *The diversity of life.* Harvard Belknap Press, Cambridge, Massachusetts.
Angier N. Feb. 28, 1995. Orangutan hybrid, bred to save species, now seen as pollutant. *The New York Times,* p. C1.
Derr M. Nov 2, 1999. Texas rescue squad comes to aid of Florida Panther. *The New York Times* p. D2.
Griffiths R. and Tiwari, B. 1995. Sex of the last wild Spix's macaw. *Nature* 375:454.
Holden C. 1997. Indians look at their big cats' genes. *Science* 278:807.
The Blue Macaws at www.bluemacaws.org
Florida Panther Net at www.panther.state.fl.us/
Orangutan Foundation International at www.orangutan.org

CHAPTER 16

XENOTRANSPLANTATION: ANIMAL ORGANS TO SAVE HUMANS

The Advisory Group on the Ethics of Xenotransplantation. 1997. *Animal tissue into humans.* Department of Health, Her Majesty's Stationery Office, United Kingdom.
Cooper D. and Ye Y. 1991. Experience with clinical heart xenotransplantation. In *Xenotransplantation: The transplantation of organs and tissues between species.* Springer-Verlag, Berlin, pp.541–557.
Chapman, L. E. 1995. Xenotransplantation and xenogeneic infections. *N. Engl. J. Med.* 333:1498–1501.
Clark M. Summer 1999. This little piggy went to market: The xenotransplantation and xenozoonoses debate. *J. Law Med. Ethics* 27:137–152.
Cozzi, E. and White, D. 1995. The generation of transgenic pigs as potential organ donors for humans. *Nat. Med.* 1:964–966.
FDA Backgrounder, September 20, 1996. Fact sheet on xenotransplantation. (www.fda.com).
Lehrman S. 1995. AIDS patient given baboon bone marrow. *Nature* 378:756.
Public Health Service. September 23, 1996. Draft public health service guidelines on infectious disease issues in xenotransplantation; notice. Federal Register. pp. 49919–49932.
Vanderpool H. Summer 1999. Commentary: A critique of Clark's frightening xenotransplantation scenario. *J. Law Med. Ethics* 27:153–157.

Weiss R. 1998. Science, medicine, and the future: Xenotransplantation. *Br. Med. J.* 317:931–932.

The Islet Foundation, Xenotransplantation Subcommittee at www.islet.org/46.htm

CHAPTER 17

CYSTIC FIBROSIS: SHOULD EVERYONE BE TESTED?

Colten, H. 1990. Screening for cystic fibrosis. *N. Engl. J. Med.* 256:328–329.

Elias S. and Annas G. 1994. Generic consent for genetic screening. *N. Engl. J. Med.* 330:1611–1613.

Kerem, B., Rommens J.M., Buchanan J.A., Markiewicz D., Cox T.K., Chakravarti A., Buchwald M., and Tsui L.C. 1989. Identification of the cystic fibrosis gene: genetic analysis. *Science* 245:1073–1080.

Livingstone J. 1993. Antenatal screening for cystic fibrosis: A trial of the couple model. *Br. Med. J.* 308:1459–1461.

Riordan, J.R., Rommens J.M., Kerem B., Alon N., Rozmahel R., Grzelczak Z., Zielenski J., Loks S., Plavsic N., Chou J.L. et al. 1989. Identification of the cystic fibrosis gene: Cloning and characterization of complementary DNA. *Science* 245: 1066–1073.

Roberts L. 1990. CF screening delayed for awhile, perhaps forever. *Science* 247: 1059–1065.

Sherif G., Brigman K., Koller B., Boucher R., and Stutts M. 1994. Cystic fibrosis heterozygote resistance to cholera toxin in the cystic fibrosis mouse model. *Science* 266:107–109.

Watson, J.D., Gilman, M., Witkowski, J., and Zoller M. 1992. *Recombinant DNA*, 2nd edition, p. 527. W.H. Freeman and Company, New York.

Wertz D. 1994. Attitudes toward abortion among parents of children with cystic fibrosis. *Am. J. Pub. Health* 81:992–996.

Wilfond B. and Fost N. 1991. The cystic fibrosis gene: Medical and social implications for heterozygote detection. *J. Am. Med. Assoc.* 263:2777–2783.

CHAPTER 18

BREAST CANCER: THE BURDEN OF KNOWING

Weinberg R. 1996. *Racing to the beginning of the road: The search for the origin of cancer.* Harmony Books, New York.

ASCO Public Issues Committee. 1996. Statement of the American Society of Clinical Oncology: Genetic testing for cancer susceptibility. *J. Clin. Oncol.* 14:1730–1736.

Deville P. 1999. BRCA1 and BRCA2 testing: weighing the demand against the benefits. *Am. J. Hum. Genet.* 64:943–948.

Ford D., Easton D., and Peto J. 1995. Estimates of the gene frequency of BRCA1 and its contribution to breast and ovarian cancer incidence. *Am. J. Hum. Genet.* 57:1457–1462.

Hartge P., Struewing J., Wacholder S., Brody L., and Tucker M. 1999. The prevalence of common BRCA1 and BRCA2 mutations among Ashkenazi jews. *Am. J. Hum. Genet.* 64:963–970.

Hoskins K., Stopfer J.E., Calzone K.A., Merajver S.D., Rebbeck T.R., Graber J.E., and Weber B.L. 1995. Assessment and counseling for women with a family history of breast cancer. A guide for clinicians. *J. Am. Med. Assoc.* 253:577–585.

Lerman C. 1996. Emotional and behavioral responses to genetic testing for suscepti- bility to cancer. *Oncology* 10:191–199.

Scientific American. September 1996. What You Need to Know About Cancer.

Struewing J., Abeliovich D., Peretz T., Avishai N., Kaback M., Collins F., and Brody L. 1995. The carrier frequency of the BRCA1 185delAG mutation is approximately 1 percent in Ashkenzi Jewish individuals. *Nat. Genet.* 11:198–200.

Wooster, R., Bignell G., Lancaster J., Swift S., Seal S., Mangion J., and Collins N. 1995. Identification of the breast cancer susceptibility gene BRCA2. *Nature* 378: 789–792.

Breast cancer information core (BIC) database at http://www.nhgri.nih.gov/Intra- mural_research/Labtransfer/Bic/

CHAPTER 19

ALZHEIMER DISEASE: ARE YOU AT HIGH RISK?

Pollen D. 1993. *Hannah's heirs: The quest for the genetic origins of Alzheimer's disease.* Oxford University Press, New York.

Post S. and Whitehouse P., eds. 1998. Genetic testing for Alzheimer disease: Ethical and clinical issues, Johns Hopkins University Press, Baltimore.

American College of Medical Genetics/American Society of Human Genetics Work- ing Group on ApoE and Alzheimer Disease. 1995. Statement on use of apolipoprotein E testing for alzheimer disease *J. Am. Med. Assoc.* 274:1627–1629.

Goate A., Charter-Harlin M., Mullan M., Brown J., Crawford F., Fidani L., Giuffra L., Haynes A., Irving N., James L, Mant R., Newton P., Rooke K., Roques P., Talbot C., Pericak-Vance M., Roses A., Williamson R., Rossor M., Owen M., and Hardy J. 1991. Segregation of a missense mutation in the amyloid precursor protein gene with familial Alzheimer's disease. *Nature* 349:704–706.

Kolata G. October 24, 1995. If tests hint Alzheimer's should a patient be told? *The New York Times,* p. 1.

Levy-Lahad E., Wijsman E., Nemens E., Anderson L., Goddard A., Weber J., Bird T., and Schellenberg G. 1995. A familial Alzheimer's disease locus on chromosome 1. *Science* 269:970–977.

Mayeux R. and Schupf N. 1995. Apolipoprotein E and Alzheimer's disease: The impli- cations of progress in molecular medicine. *Am. J. Pub. Health* 85:1280–1284.

Pollen D.A. 1993. *Hannah's heirs.* Oxford University Press, New York.

Reiman R., Caselli R., Yun L., Chen K., Bandy D., Minoshima S., Thibodeau S., and Os- borne D. 1996. Preclinical evidence of Alzheimer's disease in persons homozygous for the e4 allele for apolipoprotein. *N. Engl. J. Med.* 334:752–758.

Roses A. 1995. Apolipoprotein E genotyping in the differential diagnosis, not predic- tion, of Alzheimer's disease. *Ann. Neurol.* 38:6–14.

Schellenberg G., Bird T., Wijsman E., Orr H., Anderson L., Nemens E., White J., Bonnycastle L., Weber J., Alonso J., Potter H., Heston L., and Martin G. 1992. Genetic linkage evidence for familial Alzheimer's disease locus on chromosome 14. *Science* 2258:668–671.

Strittmatter W., Saunders A., Schmechel D., Pericak-Vance M., Enghild J., Salvesen G., and Roses A. 1993. Apolipoprotein E: High affinity binding to b-amyloid and increased frequency of type 4 allele in late-onset familial Alzheimer disease. *Proc. Nat. Acad. Sci.* 90:1977–1981.

Warrick P. Apr. 4, 1997. Nancy Reagan alone with memories for two. *Boston Globe,* p. B6.

CHAPTER 20

GENE THERAPY: THE DREAM AND THE REALITY

Annas G. and Elias S., eds. 1992. *Gene mapping: Using law and ethics as guides.* Oxford University Press, New York.

Culver K. 1994. *Gene therapy: A handbook for physicians.* Mary Ann Liebert Publishers, New York.

Office of Technology Assessment. 1984. *Human gene therapy: Background paper.* U.S. Congress, Washington, D.C.

Thompson L. 1994. *Correcting the code: Inventing the genetic cure for the human body.* Simon and Schuster, New York.

Miller A. 1992. Human gene therapy comes of age. *Nature* 357:455–460.

Mulligan R. 1993. The basic science of gene therapy. *Science* 260:926–932.

Stolberg S. November 28, 1999. The biotech death of Jesse Gelsinger. *The New York Times Magazine,* pp.137–150.

Sugarman J. 1999. Ethical considerations in leaping from bench to bedside. *Science* 285:2071–2072.

Wade N. September 30, 1999. With a death, advocates of gene therapy express concerns for future of the field.*The New York Times,* p. A20.

Walters L. 1986. The ethics of human gene therapy. *Nature* 320:225–227.

Wivel N. and Walters L. 1993. Germ-line gene modification and disease prevention; some medical and ethical perspectives. *Science* 262:533–538.

Zanjani E. and Anderson W. 1999. Prospects for in utero human gene therapy. *Science* 285:2084–2088.

National Institutes of Health, www.nih.gov/od/orda/toc.htm. Office of Biotechnology Activities, Guidelines for approval of gene therapy protocols at www4.od.nih.gov/oba/guidelines.html

CHAPTER 21

GENETIC TESTING AND PRIVACY: WHO SHOULD BE ABLE TO KNOW YOUR GENES?

Beauchamp T. and Childress J. 1994. *Principles of biomedical ethics,* 4th ed. Oxford University Press, New York.

Holtzman N. 1989. *Proceed with caution: Predicting genetic risks in the recombinant DNA era.* Johns Hopkins University, Baltimore.

Rothstein M., ed. 1997. *Genetic secrets: Protecting privacy and confidentiality in the genetic era.* Yale University Press, New Haven.

Touchette N., Holtzman N.A, Davis J.D., and Feetham S. 1997. Toward the 21st century: Incorporating genetics into primary health care. Cold Spring Harbor Laboratory Press, Cold Spring Harbor, New York.

Chapman M. 1990. Predictive testing for adult-onset genetic disease: ethical and legal implications of the use of linkage analysis for Huntington disease *Am. J. Hum. Genet.* 47:1–3.

Fears R. and Poste G. 1999. Building population genetics resources using the UK NHS. *Science* 284:267–268.

Fuller B., Ellis Kahn M-J., Barr P., Crowley E., Garber J., Mansoura M., P., Murray J., Phillips, J., Rothenberg, K., Rothstein M., Stopfer J., Swergold G., Weber B., Collins F., and Hudson K.1999. Privacy in genetics research. *Science* 285:1359–1361.

McEwen J. and Reilly P. 1992. State legislative efforts to regulate use and potential misuse of genetic information. *Am. J. Hum. Genet.* 51:637–647.

Nelkin D. and Tancredi L. 1991. Classify and control: Genetic information in the schools. *Am. J. Law Med.* 17:51–75.

Wertz D., Fanos J., and Reilly P. 1994. Genetic testing for children and adolescents: Who decides? *J. Am. Med. Assoc.* 272:875–881.

CHAPTER 22

FROZEN EMBRYOS: PEOPLE OR PROPERTY?

Human Fertilisation Embryology Authority. 1995. *Code of Practice.* London.

Radin M. 1996. *Contested commodities.* Harvard University Press, Cambridge, Massachusetts.

Robertson J. 1994. *Children of choice; Freedom and the new reproductive technologies.* Princeton University Press, New Jersey.

American Fertility Society. 1994. American law and assisted reproductive technologies. *Fert. Steril.* 62:8S–12S.

Bonnicksen A. 1992. Genetic diagnosis of human embryos. *Hastings Centr. Rep.* 22: S5–S11.

Burn J. and Strachan, T. 1995. Human embryo use in developmental research. *Nat. Genet.* 11:3–6.

Committee of Inquiry into Human Fertilisation and Embryology. July 28, 1984. Recommendations of the Warnock Committee. *Lancet* 2:217–218.

The Ethics Committee of the American Fertility Society. 1990. Ethical considerations of the new reproductive technologies. *Fertil. Steril.* 53:1S–104S.

Goldberg C. November 5, 1999. Massachusetts case is latest to ask court to decide fate of frozen embryos. *The New York Times,* p. A19.

Margolick D. June 27, 1984. Legal rights of embryos. *The New York Times,* p.17.

News.1995. Embryo research tests NIH's mettle. *Nat. Med.* 1:5.

Schwarz J. September 28, 1994. Panel backs funding of embryo research; abortion foes denounce proposed rules. *The Washington Post*, p. A1.

Simpson J. and Carson S. 1992. Preimplantation genetic diagnosis. *N. Engl. J. Med.* 327:951–953.

Yardley, J. April 17, 1999. Investigators say embryologist knew he erred in egg mix-up. *The New York Times*, p. A13.

CHAPTER 23

CLONING: WHY IS EVERYONE OPPOSED?

Di Berardino, M. 1997. *Genomic potential of differentiated cells.* Columbia University Press, New York.

Di Berardino, M. 1998. *Cloning: Past, present, and the exciting future.* Federation of American Societies for Experimental Biology. Breakthroughs in Bioscience Series (www.faseb.org/opar/cloning).

Report and Recommendations of the National Bioethics Advisory Commission 1997. *Cloning human beings,* Rockville, Maryland.

Editorial. 1997. Clone encounters. *Nat. Genet.* 15:323–324.

Kato Y., Tani T., Sotomaru Y., Kurokawa K., Kato J., Doguchi H., Yasue H., and Tsunoda Y. 1998. Eight calves cloned from somatic cells of a single adult. *Science* 282: 2095–2098.

Nature, vol. 385. February 27, 1997. Front cover.

Robertson J. 1999. Two models of human cloning. *Hofstra Law Rev.* 27:609–639.

Robertson J. 1999. Ethics and policy in embryonic stem cell research. *Kennedy Inst. Ethics J.* 9:109–136.

Shapiro H. 1999. Ethical dilemmas and stem cell research. *Science* 285:2065.

Vogel G. 1999. Capturing the promise of youth. *Science* 286:2238–2239.

Wakayama T., Perry A., Zuccotti S., Johnson K., and Yanagimachi R. 1998. Full-term development of mice from enucleated oocytes injected with cumulus cell nuclei. *Nature* 394:369–374.

Wilmut I., Schnieke A., Mcwhir J., Kind A., and Campbell K. 1997. Viable offspring derived from fetal and adult mammalian cells. *Nature* 385:810–813.

The Joint Steering Committee for Public Policy at www. jscpp.org/jscpp/Stemcell.htm

Pharmaceutical Research and Manufacturers of America (PhRMA) at www. phrma.org/genomics/cloning/stem.html

CHAPTER 24

EUGENICS: CAN WE IMPROVE THE GENE POOL?

Kevles, D. 1985. *In the name of eugenics.* Knopf. New York.

Ludmerer K. 1972. *Genetics and American Society,* Johns Hopkins University Press, Baltimore.

Osborn F. 1968. *The future of human heredity: An introduction to eugenics in modern society.* Weybright and Talley, New York.

Pearson K. 1924. *The life, letters and labours of Francis Galton,* Volume II. *Researches of middle life.* Cambridge at the University Press, United Kingdom.

Reilly, P. 1990. *The surgical solution: A history of involuntary sterilization in the United States.* Johns Hopkins University Press, Baltimore.

Board of Directors of the American Society of Human Genetics. 1999. Eugenics and the misuse of genetic information to restrict reproductive freedom. *Am. J. Hum. Genet.* 64:335–338.

Editorial, September 12, 1993. Wanted: Smart sperm. *Boston Globe,* pp.1, 39.

Gill M. March, 1995. The genius babies. *Ladies Home Journal* pp.76–82.

Horgan J. June 1993. Eugenics revisited. *Sci. Am.* 268:122–131.

Index